INFORMATION RESOURCES SERIES

Guide to Basic Information Sources

in

ENGINEERING

INFORMATION RESOURCES SERIES

Planning Editor: The Late Bill M. Woods

Guide to Basic Information Sources in
ENGINEERING

by
Ellis Mount
Science Bibliographer
Columbia University Libraries

A Halsted Press Book
by Jeffrey Norton Publishers, Inc.

John Wiley & Sons
New York   London   Sydney   Toronto

Library of Congress Cataloging in Publication Data

Mount, Ellis

Guide to basic information sources in engineering.

(Information resources series)

1. Technical literature. 2. Engineering—Bibliography. 3. Technology—Information Services. I. Title.

T10.7.M68 1976   607   75-43261
ISBN 0-470-15013-0

Copyright © 1976 by Jeffrey Norton Publishers, Inc. All rights reserved. Printed in the United States of America. No part of this publication may be reproduced, stored in a retrieval system, or transmitted, in any form or by any means, electronic, mechanical, photocopying, recording, or otherwise, without the prior written permission of the publisher.

Distributed by Halsted Press,
A Division of John Wiley & Sons, New York.

Preface

Technical information is a precious but semiperishable commodity. It is precious because recorded information is the basis for all subsequent progress and accomplishments in engineering and other sciences—the present-day practitioner in these fields stands on the shoulders of those who went before, profiting from their recorded experience. It is semiperishable because it quickly becomes outdated, largely because of improved techniques and discoveries that the information itself brings about. In turn, new information is published. Obviously, some theoretical information remains valid for a long time, but the rest is subject to constant change.

It is precisely because the engineer, at whatever level, needs to keep up with this constant change, and needs to know **how** to keep up with it with a minimum of time and effort, that I have written this book. It emphasizes which type of literature should be used for each particular purpose, and stresses the features and advantages—as well as the limitations and disadvantages—of more than a score of types of technical literature, and also supplies outstanding examples of each.

This guide is intended primarily for engineering students and researchers. It is a source book of information about engineering and allied fields, and in particular about how to find this information in the reference room or information center once those at the reference desk have been queried and the card catalog consulted.

I have made no attempt to compile an exhaustive listing of all known works in any category. Such a listing would be enormous and entirely impractical for the engineers for whom this book is designed. By concentrating on some of the better references, I hope that engineers will be led to explore further on their own, should they need additional sources.

In all but a few—specified—cases, only books still in print are listed, so that those who wish to build a personal reference library can more easily do so.

Although the emphasis is naturally on technical literature, some attention is also given to a few important general works or works in allied fields. The examples in the longer chapters are broken down into subsections, according to the field of learning, and listed alphabetically thereafter. Cross-references are given in the Index.

Much of this book is based on a one-semester course in technical literature that I have given for several years at the School of Engineering and Applied Science at Columbia University. The goal of the book is the one that also pervades that course—to make engineering students aware of the need to use technical literature and to give them the techniques and understanding for using it well. I hope that this volume will be useful for those who want to improve their knowledge of this subject by self-instruction, as well as possibly serving as a text for formal courses in technical literature. I hope, also, that technical libraries will find it a worthwhile addition to their reference shelves.

I wish to express appreciation for the inspiration evoked by the inquiring minds of my students, and for the assistance given me by many of my colleagues, particularly the staffs of three libraries—the Engineering Societies Library, The Science and Technology Research Center of The New York Public Library, and the science units of Columbia University Libraries.

Ellis Mount

Contents

Preface, i
Note to the Beginning Researcher, 1

PART I. TECHNICAL LITERATURE: WHAT IT IS, WHERE TO FIND IT, 7
  1. The Nature of Technical Information, 9
  2. The Role of Libraries and Information Analysis Centers, 13

PART II: BOOKS, 21
  3. Bibliographies, 23
  4. Dictionaries, 29
  5. Encyclopedias, 43
  6. Handbooks, 49
  7. Guides to the Literature, 79
  8. Histories, 85
  9. Biographical Information, 87
  10. Directories and Yearbooks, 91
  11. Annual Review Series, 97

PART III: PERIODICALS AND TECHNICAL REPORTS
  12. Periodicals, 103
  13. Periodical Indexing and Abstracting Services, 107
  14. Technical Reports, 119
  15. Technical Report Indexing and Abstracting, 121

PART IV: OTHER SOURCES OF INFORMATION
  16. Conferences and Symposia, 129
  17. Specifications and Standards, 133
  18. Manufacturers' Catalogs & Directories of Manufacturers, 137

19. Tables, Statistics, and Data Compilations, 141
20. Patents and Trademarks, 153
21. Dissertations and Masters' Essays, 157
22. Newspapers, 159
23. Translations, 161
24. Maps, Atlases, and Engineering Drawings, 163
25. Business Information, 167
26. Quotations Indexes and Thesauri, 171
27. Personal Contacts, Including Professional, Industrial, Educational and Governmental Sources of Information, 173
28. Special Services and Materials (Current Awareness Services; Retrospective Searching; Electronic Data Processing), 177

Index, 183

Note to the Beginning Researcher

LIBRARY RESOURCES

When people are seriously interested in a subject and need more information they frequently head for a library. However, at the doorway, confusion often sets in owing to uncertainty about how best to use the library's resources. The purpose of this chapter, therefore, is to provide basic information to the beginning researcher on how to get the most out of time spent in the library.

The Librarian

First, it is a good idea to make use of the knowledge and expertise of the reference librarian, who may well turn out to be your guide not only at the beginning of your library work but throughout your professional career. Asking your librarian about the subject you are researching may well bring suggestions, new ideas or approaches, and other unexpected bonuses. As an example of the last mentioned, many pamphlets, circulars, and other materials are not listed in the card catalog but are kept in file cabinets or on shelves behind the reference desk.

The Card Catalog

Once oriented, most researchers first go to the card catalog, that modest to imposingly large collection of file drawers that contain arrays of several thousand or even millions of 3 x 5 cards. The catalog contains on its cards an alphabetical listing of all the publications in that library. All books are listed, or indexed, by the author's name. Usually they are also indexed by title and subject. Thus, a book such as the **Rare Metals Handbook**, edited by Clifford A. Hampel, will be listed under "Hampel, Clifford A." and under the title as well. If the library has a thorough card catalog this book will also be listed under a subject, such as "Metals, Rare."

Many libraries have one catalog only, with all types of listings in it, alphabetized from A to Z. Others have three catalogs, one each for author, title, and subject. In addition, some libraries have separate sections of their catalogs reserved for periodicals, or serials, listings. (**Serial** means any publication that is put out in a **series**, such as most magazines, some series of books that are published a volume at a time, or books such as the yearly **Who's Who**.) There are also cross-references that tell the user to **see** or **see also** another entry for related materials.

The card catalog will almost always give the researcher a starting point. It lists subject headings for the contents of each publication, as well as other basic information, such as edition, publisher's name, and date of publication. The call number, in the upper left corner of the card, both identifies the book and tells where it is in the library.

If the "open shelf" system is used, the shelves are open to all and most of the books in the library are obtained by self-service. Either the librarian or a printed diagram will direct the researcher to the shelf or alcove where the books are filed in call-number sequence. If the "closed system" is used, the researcher fills out a call slip for the book, writing the call number on it, and presents this slip at the circulation desk. One of the library staff will then get the book. See page 16 for a more detailed discussion of the use and understanding of the card catalog.

The Reference Room

Although the card catalog is the starting point in the library, the reference room is its real heart. It is there that most of the basic volumes of considerable importance are found. Books in the reference room may not be taken from the library. They must be used only in the reference section, the main reading room, or occasionally in other designated areas. This is because the reference room contains indispensable research volumes that are in constant demand and must always be on hand. Fundamental among these volumes are the indexes to articles in magazines, journals, and other sources. These indexes frequently provide the only clues to lead the researcher forward.

NOTE TO THE BEGINNING RESEARCHER

There are also such references as **Books in Print** which lists books still obtainable—that is, still in print—even though they may not necessarily be available in a particular library. If a book is **not** in print, the researcher will be able to find it only in a library or a second-hand book store.

And A Note about This Book

One major point about this book is that the index should be freely consulted. If a specific author, work, or subject is in question, the index will be the quickest way to find all references to it. Cross-references also are indicated in the index.

FINDING SOURCES OF INFORMATION

Although locating information efficiently is not an exact science, there are a few basic rules that are helpful to follow. A number of points that are good to bear in mind when you're making a search are listed below.

When you are working alone:

1. **Determine clearly what you need.** If you don't know exactly what you're looking for, you'll have to take the time to get a general overview, then narrow it down. Of course, it's always possible that such an overview will turn your search around completely, but until that happens make sure you have a definite goal.

2. **Plan your schedule generously.** Allow more time than you think you need. Your search may take longer than you think. There are a lot of data out there!

3. **Determine what you already know.** Make a list of likely agencies, authors, or periodicals that might be of help and take it with you. The author's name, or the names of companies and government agencies working in the field involved, might be in that article you just read. Find dates, if possible. There's no point in searching for theoretical articles about lasers in 1950 periodicals indexes since that would be about five years too early.

4. **When appropriate, use the easiest tools first.** In general, certain types of books, such as dictionaries, technical encyclopedias, and handbooks, are easier to work with and to get at

physically than are patents, technical reports, or periodical articles. However, if you know for sure that what you want can be found only in a patent or in a technical report, then using that source first is of course fastest.

5. **Be specific in your approach unless experience calls for a more general approach.** When you look up a subject, try to find the most specific term for it that you can. If you're looking for information on "hydrochloric acid production" then look that up, not the broader term of "acids." In some sources, such as **Chemical Abstracts**, it is a complete waste of time to look up the broader term, and, in addition, it's best to go directly to a subheading indicating "preparation" or "manufacturing" or whatever aspect you want. Especially when you're using the card catalog, **be specific**.

Conversely, however, if your source does not include your specific term, you may still find information in a work on a broader topic. A book on chemical engineering or industrial chemicals may have a chapter on the production of hydrochloric acid. Or, to use another example, although no one would expect to find an entire book devoted to the design of landing gears for jet planes, certainly there would be patents, periodical articles, and technical reports on this subject.

6. **Keep an open mind about likely sources.** Do not get so absorbed in a 10-year-old book that you forget the many newer sources on your subject that you should also check. Or, conversely, do not feel that just because an article is from a current issue of a journal there is no need to search further.

7. **Avoid nonproductive searching.** Do not give up too easily, but also avoid spending half a day in the early stages of a search when you're getting nowhere. Perhaps you're using the wrong terminology, time period, or type of literature. Rethink, or ask for help.

8. **Ask for help when you need it.** Most libraries and information centers have capable staff members who are hired specifically to help you find information. When you need their help, ask for it!

When you need the librarian's help:

1. **State your need clearly.** Obviously, unless the librarian knows exactly what you need, he or she can't possibly do you much good.

NOTE TO THE BEGINNING RESEARCHER

One reason the librarian is there is to help you, but clarity on your part will bring you better, faster help.

2. **State what you have already done.** Stating briefly what you already know can save much time in getting you on the right path.

3. **Indicate your time limits.** Perhaps in three weeks the very item you wanted will come along—still in time for your deadline? Or too late? State how long you are apt to be interested, or what your absolute deadline is.

Many other common sense points might be listed, but these are among the most important ones. Some of the following examples, which illustrate our main points, are actual cases, and others could well have been.

Case 1. **Vaguely stated needs.** An engineer asked for help in finding information about radar circuits. When shown several books on radar, he showed no interest. Finally it came out that his real interest was in circuits used in aircraft instrument-landing systems that used radar. This called for a completely different approach, and the reference librarian was able to direct him to several periodical and technical report indexing services where he found references to what he needed.

Case 2. **Ingenuity in the choice of a source.** Another person, who was looking for a definition of "tribology," found no technical dictionary or encyclopedia that included the term. Deciding that it must therefore be relatively new, the person turned to a periodical indexing service which led to a periodical that contained the proceedings of a 1976 conference devoted to tribology, and which included a brief statement about the origin of the term. It might have been years before a book or a dictionary had included this term.

Case 3. **Improper use of a source.** An engineer looking for a quick description of a water-activated battery tried a multivolume technical encyclopedia, looking under all the proper terms, and found nothing. The reference librarian suggested they consult the index to the set, when they found the term described in an article that otherwise would have been overlooked. Indexes and tables of contents are almost always well worth consulting.

Case 4. **Poor choice of reference sources (too old or too complicated).** An engineer seeking information on mercury pollution of streams and lakes consulted a good handbook and a good encyclopedia but found nothing. It was then pointed out that the subject had just come into general study in the past two or three years, and that both sources had been published before that. Once again periodical indexes covering chemistry, biology, and environmental engineering led to many references on the subject.

Another engineer asked for help in using the index to U.S. patents, including the classification manual. An hour later the search had proved fruitless. It developed that what was wanted was very basic information about the use of computers in controlling machine tools, such as lathes, in mass production work, for which the patent index was a totally unsuitable information source. Instead, a book on the subject as well as a chapter in a handbook on data processing systems were suggested as references. Patents are essentially for only the most advanced specialists in a field.

An excellent source on the methodology of searching technical literature is listed below.

American Chemical Society. **Searching the Chemical Literature.** Advances in Chemistry Series no. 30. Columbus, Ohio: 1961. 326 pp.

> This carefully prepared volume indexes papers ranging from specific topics (house organs and trade publications, or chemical literature of Germany) to more general ones (library techniques in searching, language problems in literature searching, etc.).

Part I
Technical Literature: What It Is, Where To Find It

The first chapter of Part I describes the basic characteristics of technical literature, including its significance to engineers and their work. The second chapter describes the general role of the library and the information analysis center in serving engineers. It also discusses the techniques and principles involved in making successful searches for information.

CHAPTER 1

The Nature of Technical Information

The engineering student and the practicing engineer live in a world characterized by two major forces that produce a technical climate far different from that of even half a century ago. One force is the greatly increased rate of change of technology and engineering practice: today's standard method is tomorrow's outdated mode. This is illustrated by the decreasing time lag between the theoretical study and its transformation into a relatively common, commercially made product. For example, the laser was a delicate, very expensive laboratory instrument just a few years ago, with only sparse references to it in the literature. Then, just a few years later, it was being made on the production line, and it is now cheap and rugged enough for even hip-pocket field models to be readily available. And thousands of descriptions of lasers are available in all types of literature.

The second force affecting technical developments is the greatly increased complexity of our society. It is becoming rare for a technical person to work exclusively in one discipline now, at least as compared to several decades ago. The boundary lines between different branches of science and engineering are becoming ever more blurred. For instance, today we find physicists becoming interested in the relation of physics to biology, as in the space program, and ocean engineers combining many talents, including oceanography, structural engineering, and electronics. Also, technology has become so important to society that each technical project must be planned with its social and economic aspects clearly in mind.

These two forces are reflected in the technical literature, rapidly superseding material that has only rather recently been published. The literature itself then helps to increase the rate of change in science and engineering, which in turn creates more literature—a self-feeding system. Derek J. de Solla Price writes that the number of technical periodicals has been growing at an exponential rate in the last two centuries. In 1750 there were approximately 10 scientific periodicals. This figure increased to 1,000 titles by 1850, then to around 80,000 by 1950.[1] Price points out that the number of periodical abstracting journals is following a similar pattern and is increasing by a factor of 10 every 50 years.

Another example of this "information explosion" is the number of articles indexed annually by **Chemical Abstracts**, the chief source for abstracts in its field. In 1960 it abstracted 145,000 items per year. This figure rose to around 300,000 items by 1970, and to half a million by the mid-1970s. The rate of increase of papers from 1960-1970 was about 8.4 percent per year, **or double the total amount every nine years**. The annual rate of increase for patents covered by **Chemical Abstracts** is 10.9 percent.[2] Clearly, the literature is expanding so fast that no engineer can keep up with it.

The increased blurring of boundary lines between disciplines is also evident in the literature. For example, journals from civil engineering societies may include articles on the effects of urban planning in areas where skyscrapers are to be built on land previously considered unsafe because of earthquakes. Today, such articles must consider geology, economics, advanced structural design, urban politics, and so forth, in making recommendations. Articles on aircraft design must consider political, medical, psychological, atmospheric, and economic factors. Of course some studies report on the technical aspects alone, but the wise engineering firm no longer stops there.

There is also a growing number of **types** of information forms. No longer can printed matter alone be used. Today, in addition, there are audio and video recordings, microfilm, and computer-based information files.

---

[1] Derek J. de Solla Price, **Science Since Babylon**. 2d ed. New Haven, Yale University Press, 1975. 216 pp.

[2] Dale B. Baker, "World's Chemical Literature Continues to Expand." **Chemical and Engineering News** 29 (**28**): 37-40, 12 July 1971.

THE NATURE OF TECHNICAL INFORMATION

Quite often engineering students complete their schooling with only a meager idea of what is available in the technical literature. Indeed, many faculty members are hardly better off and thus do little to guide their students in this area. Specialists frequently develop a useful coterie of colleagues in their subject fields, often on an international basis, who keep each other informed of major events and progress. Even so, this procedure is neither infallible nor all-inclusive, and it is certainly beyond access by most engineers, not to mention the beginning ones.

And engineers do need to keep up with the literature, for in a few years much of what is current will be outdated. Also, one must know how to check or recheck facts, or how to learn new fields as necessity demands.

I have two goals for this book. The first is to show the **wealth** of information in the literature. Sources range from news items of current interest to the more enduring, basic articles, monographs, and texts. The second goal is to indicate the good and bad points of over 25 different **types** of material and to show not only when each type should be used but also how to locate it.

I have made no attempt to list all the available examples of literature of any type. To do so would be far beyond the scope of this book. The sources chosen are among the best examples available, and are works with which an engineer should be familiar. Annotations describe major features and their uses, and hopefully will spur readers to seek out other works that meet their particular needs.

CHAPTER 2

The Role of Libraries and Information Analysis Centers

Types of libraries range tremendously. There are tiny public libraries with virtually no technical literature as well as large libraries that have hundreds of thousands of volumes of technical material. Library staffs also vary greatly. Some have almost no one able to help engineers with technical problems, and others have specialists with advanced degrees in science or engineering as well as graduate library degrees.

Generally speaking, the most technically trained staff members and the most concentrated collections of technical literature are to be found in (a) the science departments of the largest public libraries, (b) sections of university libraries devoted to science and engineering, or (c) the so-called special libraries devoted exclusively to technology and science. Such special libraries may be operated primarily for employees of a particular industrial firm or for a government research firm or similar nonpublic organization. Special libraries tend to be very complete in their subject specialties, often including many little-known items beyond the ability of the all-encompassing technical libraries to acquire. On the other hand, special libraries are often hard-pressed for space and do not try to maintain large collections of older literature. For this sort of material, the university and large public library collections are more useful. It would be a rare special library that had scientific journals published as far back as the 1800s, yet good university libraries often keep complete runs of important journals, as do many of the largest public library science departments. Older books are also commonly retained on the same pattern as older journals.

Engineers should expect their employers to furnish some technical literature. Many employers maintain excellent technical libraries and offer more services than a university or public library. Some services offered include automatically sending engineers items of current interest to match their particular fields of interest, doing extensive literature searches, ordering obscure documents, or locating specific data. All of this is far beyond the capability of public and college libraries, which generally cannot afford the staff necessary for such service.

The library that does not have a wanted item of literature can usually borrow it on interlibrary loan from another library. In cases where that library does not wish to loan the item, perhaps because it is out of print and cannot be replaced if it is lost, making a photocopy is sometimes the answer. With longer items, where a photocopy would be too expensive, microfilm copies are often available.

Sometimes networks of cooperating public and college libraries issue catalogs of their joint periodical or book holdings (called "union lists") which list the materials held by each library. Sometimes the networks are linked together by telephone, teletype, or even facsimile equipment suitable for transmitting documents immediately over long distances. Sometimes computers are used to maintain the data base—that is, the information available in the network.

Most libraries are included on one or more network lists of libraries of like types, such as public, school, or special. Librarians usually have access to at least the most widely held of these directories and can often direct inquirers to a particular library which may let them use special material on its premises. Sometimes this is the only way unique material can be obtained.

INFORMATION ANALYSIS CENTERS

World War II spawned a variation and extension of the special library which is now commonly known as the "information analysis center." Such centers are similar to libraries in that both have publication and document collections and both offer services to users; however, they vary tremendously in the services offered, in staffing, and in details of organization. Usually they are limited in scope to a single specialized subject area.

ROLES OF LIBRARIES AND INFORMATION ANALYSIS CENTERS 15

Information analysis centers usually serve a select clientele, much as most industrial special libraries do. But their services go beyond what many libraries are able to provide. The principle difference is that an information analysis center is concerned with **evaluating** literature, and often issues summaries and surveys describing the relative value of literature on a particular subject. Rather than merely locating some data for a requester, they can usually go a step further and offer comments on the comparative merits of the information.

To do this in a professional manner, information analysis centers usually have technical staff members who have had considerable experience and/or education in the field involved. Some centers also have staff members who can analyze foreign-language material. Needless to say, such centers are expensive to operate, and often government agencies and very large companies are the only sponsors who can afford to maintain them.

The subject matter of interest to the centers is, as mentioned, often quite limited, and depends on the subject of interest to the group involved. It might be, for example, nuclear reactor control devices, or a particular metal, or a particular type of test, such as thermophysical data. Those centers sponsored by the government often make some effort at serving the public, but charges for the services are common. One of the main problems in using the services lies in just finding out if an information center exists on a particular subject. Again, most librarians have access to lists that give names and locations of major centers.

USING LIBRARIES

This section is devoted to helping the engineer become familiar with the way libraries are usually arranged and with that often little-understood tool, the card catalog. The subject is here dealt with in far greater detail than in the preceding Note to the Beginning Researcher. In addition, what applies here may vary considerably from the way in which information analysis centers are arranged, since the centers are usually set up for direct use by the staff, whereas most libraries are arranged to encourage self-use by patrons, with the staff on hand to help only if needed.

THE LIBRARY CARD CATALOG

Although most researchers have some idea of how to use libraries efficiently, few are skilled at it. Many do not even know how to use the card catalog, that venerable tool in use since the turn of the century. Therefore, a few words on the subject.

The most common tool the library has is the traditional card catalog. In some libraries and some information centers this catalog is being replaced wholly or in part with computer-printed indexes, usually called book catalogs, since they are in the form of a printed list rather than a card file. However, the two variations are so closely allied that information given here will generally apply to both.

The card catalog usually includes cards for the following:

1. Authors, titles, subjects, and series numbers for **books**.

2. Titles and general broad subjects of **periodicals** (for the entire set of a periodical, nor for individual articles).

3. Titles and broad subjects for sets of other types of serials such as annuals, yearbooks, "Advances in" series, certain report series, etc.

4. Individual pieces of **unbound material**, such as important specific technical reports or pamphlets. Usually the same type of cards used for books are used here.

The card catalog does **not** supply information about individual periodical articles, chapters in books (in most cases), or individual technical reports (in most cases). The smaller the library's scope, the more likely its catalog is to contain special cards, called "analytics," which describe specific portions of larger works.

The cards are usually arranged in one alphabet, called a dictionary catalog. Occasionally subject cards are in a separate file. Book catalogs frequently have separate volumes for authors, titles, and subjects so as to reduce the size of any one volume, and are often printed as an annual or a five-year cumulation.

Thus, although the card catalog furnishes very basic information, it needs to be supplemented by other tools, such as bibliographies, periodical and report indexes, etc., which will be discussed later.

Arrangement of the Collections

Most libraries arrange their books by either the Dewey decimal system or the Library of Congress system. Both are based on a list of subjects arranged in a specific, logical system. University libraries now generally use the Library of Congress system, while smaller college libraries and most public libraries use the Dewey system. In a few rare instances some other system is used. The object of all classification systems is to arrange books by like subjects, then alphabetically by author within a given subject.

The Dewey Decimal System (named after its inventor, Melvil Dewey,) uses numbers only, with all knowledge arranged from numbers 000 to 999. For example, engineering subjects are put in whole-number classifications, which are represented by numbers. Engineering itself is found under 621, but electronics is represented by 621.338, a subdivision of the larger class of 621.3, which includes all electrical engineering.

The Library of Congress system uses a combination of letters and numbers. Here engineering is found under T, electrical engineering is under TK, while a subdivision for transistors is represented by TK followed by both numbers and letters.

To each of the classification numbers used in either system the cataloger adds a special code for every individual book in that class. This code is called the book number and is part of the call number, which includes the entire set of characters used to identify a book. The book number is usually on the last line, and begins with the first letter of the last name of the author given credit for the book, whether a personal author or a group (called a corporate author). A book on transistors by an author named Romano might have the book number:

    TK
    7872
    T73
    R75

where the first three lines are the class number and R75 is the book number. If a publication date is shown for an edition other than the first, it may follow the book number.

Filing books by their call numbers has two results. First, like subjects are filed together by class number. Second, all books on a given subject are then filed alphabetically by author, or by title if no author is named. This system makes browsing feasible when one wants to see all the books on a given subject, yet it also keeps together all the books by a given author on that subject. Those few libraries whose stacks are not open to the public may file books by size, order of acquisition, etc.

Some libraries arrange periodicals by their call numbers, and thus by subject, while others file them alphabetically by title.

Technical reports, pamphlets, and other small items may be filed in cabinets (often called vertical files) or on special shelving. They are usually filed by number, which places them in an apparently random order and makes browsing difficult. One must usually rely on special indexes or the reference librarian to locate these items.

CATALOG CARDS

A catalog card gives a considerable amount of information about a book. Besides the title, author, and publisher, it shows the publication date, certainly of prime importance with technical subjects, which may soon get out of date. The paging can be a practical measure of the extent of a work, and any special illustrative material, such as colored plates, which is often extremely important in subjects like metallography and biology, is indicated following the paging. If a work is best known as part of a series, this is also shown.

The subject headings assigned to a work, shown toward the bottom of the card with Arabic numerals, indicate the main thrust of a book. Joint authors or names of sponsoring agencies are also shown (preceded by Roman numerals). All these elements are called the **tracings**, and there will be a card in the catalog for each of them.

The card also shows the call number assigned by the Library of Congress as well as the equivalent number in the Dewey decimal system.

Most libraries buy their catalog cards from the Library of Congress. However, that agency now issues a weekly reel of magnetic tape containing all the information found on the average catalog card, with more and more of its cataloging output available in this format. This is known as the MARC (Machine Readable Catalog) system. Cooperative library networks, such as that operated by the Ohio College Library Center, also provide libraries with cataloging information by means of a computerized file.

Part II
Books

This section concerns information that commonly appears in books. It includes such book types as technical dictionaries and encyclopedias, handbooks, histories, and so forth. These standard reference tools tend to give more general information than does the literature cited in the following parts.

CHAPTER 3

Bibliographies

Bibliographies are listings of literature that have been selected and organized for a particular purpose. They come in many forms—as, for example, separate books or multiple volumes of books; separate pamphlets, especially on newer subjects; special issues of periodicals; lists at the ends of books, chapters, or journal articles.

General bibliographies are those that include information useful to almost every type of reader. Usually they treat books rather than periodical literature. The major commercial bibliographies are typical of these compilations. They list all United States literature—generally either all that was printed in a given year, or all that is still in print, regardless of publication date. They cite works by author, title, and subject, and can thus be a guide to information in these areas as well as to possible books to consult on a multitude of subjects.

Specific bibliographies focus on precise subjects. They may be restricted in any of a number of ways: to a particular field (probably the most common limitation), to the works of a certain author or group of authors, to a certain time period, or to a particular type of literature. Or, a bibliography of American patents issued during 1965-1968 on the welding of steel would obviously combine many of these limitations and be restricted in ways of its own.

The best bibliographies of any kind have detailed subject indexes. They also indicate their standards of inclusion and list the sources of literature used—such as, only the major articles found in the last ten years of the eight top electronics journals. In addition, they provide some description of the contents of each entry so that the reader can better tell whether to check it further. These descriptions usually fall into one of two broad classes: an **indicative abstract**, which indicates only the highlights of an entry, or an **informative abstract**, which gives a much more detailed and informative picture—and in which, on occasion, one can find the important particulars without even going to the original writing.

Most card catalogs as well as periodical indexing services have a subheading under subject headings to indicate bibliographies, such as "Mining—Bibliographies."

Needless to say, finding a bibliography of the type one wants can save countless hours of searching.

SELECTED EXAMPLES

General

American Book Publishing Record. New York: Bowker, 1960- [Monthly].

> This is a listing, arranged by Dewey decimal classification, of new American books on all subjects. It gives full bibliographic citations, including assigned subject headings, but no annotations. Each issue has its own author and subject-title indexes. A separate annual cumulation is also available.

Books in Print: An Author-Title-Series Index to the "Publishers' Trade List Annual." New York: Bowker, 1948-   [Annual] 2 vols.

> This bibliography lists books currently available from American publishers (some 400,000 titles from some 3,500 publishers) and is arranged alphabetically by author in one volume and by title in the other. Subject research may be pursued in a companion publication, **Subject Guide to Books in Print**. **Books in Print** is a valuable reference source found in almost all libraries and is updated by **Forthcoming Books**. Since 1972 a supplement volume has been issued each spring.

Cumulative Book Index. New York: Wilson, 1898-   [Monthly]

> The **CBI** lists newly published books in the English language, with authors, titles, and subjects all arranged in one alphabetical index. It cumulates quarterly and annually but has had larger cumulations in the past.

Forthcoming Books. New York: Bowker, 1966-   [Bimonthly]

This periodical updates **Books in Print**. It lists books that will appear within the next five months as well as those published since the last **Books in Print**. There are separate indexes for authors and titles but not for subjects. It is supplemented by **Subject Guide to Forthcoming Books**.

Paperbound Books in Print. New York: Bowker, 1955- [Semiannual]

The work lists all paperback books in three indexes—author, title, and subject—and is very useful for those seeking paperback editions.
The 1976 edition contained some 132,000 titles.

Subject Guide to Books in Print. New York: Bowker, 1957- [Annual]

This is a listing by subject of most of the items in **Books in Print** and is thus restricted essentially to English-language books. Some books appear under more than one subject heading. This work is a very handy source for quickly locating books on a given subject and lists more than 380,000 books with more than half a million entries under some 60,000 headings. It gives bibliographic data as well as price and is found in almost all United States libraries of any size.

Subject Guide to Forthcoming Books. New York: Bowker, 1974- [Bimonthly]

Arranged by subject, the guide supplements **Forthcoming Books**.

Weekly Record. New York: Bowker. 1974-   [Weekly]

This listing of newly published books, arranged alphabetically, was formerly included in **Publishers Weekly**.

General—Science and Engineering

Jenkins, Frances B., comp. **Science Reference Sources.** 5th ed. Cambridge, Mass.: M.I.T., 1969. 231 pp.

This bibliography of titles is typically found in large reference libraries. It is arranged by type of literature, under the broad categories of science, mathematics, physics, chemistry, astronomy, earth sciences, biological sciences, medical sciences, agricultural sciences, and engineering. It has an author-title index but no annotations.

McGraw-Hill Basic Bibliography of Science and Technology. Edited by T. C. Hines. New York: McGraw-Hill, 1966. 738 pp.

Using an alphabetical arrangement, this bibliography cites one or more briefly annotated references for over 7,000 technical and scientific terms. An appendix organizes all the terms into around 100 broad classes.

New Technical Books: A Selective List with Descriptive Annotations. New York: New York Public Library, Research Libraries, 1915- [Ten issues annually]

The selective list of new books recently added to the collection of the Libraries' Science and Technology Division gives the contents of as well as brief annotations for each book. Arranged by subject, each issue has an author and subject index, which cumulates annually. The work presents a broad, comprehensive array of new literature.

Technical Book Review Index. Pittsburgh: Special Libraries Association, 1935- [Ten issues annually]

This selected compilation of book reviews from periodicals describes new books on engineering and the other scientific disciplines. The work presents a well-balanced group of books and has an annual author index.

Chemical Engineering

Bourton, Kay., comp. **Chemical and Process Engineering: Unit Operations, a Bibliographic Guide.** Washington, D.C.: IFI/Plenum, 1968. 534 pp.

The bibliography of thousands of important books, technical reports, and periodical articles on the subject is arranged under 38 headings such as ion exchange, flotation, and ultrasonics. All items are annotated.

Computers and Data Processing

U.S. National Bureau of Standards. **Computer Literature Bibliography.** Compiled by W. W. Youden. Washington, D.C.: Superintendent of Documents, 1965, 1968. 2 vols. (NBS Miscellaneous Publication Nos. 266 and 309.)

Volume 1 covers literature published from 1946 to 1963, and volume 2 is for 1964-1967. The bibliography uses a Key Word In Context (KWIC) index format for titles and has an author index. There are no annotations.

Environmental Studies

Winton, Harry N. M., comp. **Man and the Environment: A Bibliography of Selected Publications of the United Nations System, 1946-1971.** New York: Unipub, 1972. 305 pp.

This volume lists nearly 1,200 items, arranged by broad subjects, covering a wide assortment of topics. Most items are annotated. There are indexes for subjects, titles, authors, and series numbers.

Materials

Johnson, H. Thayne, ed. **Electronic Properties of Materials: A Guide to the Literature.** New York: Plenum, 1965. 2 vols.

This is essentially an elaborate bibliography. Volume 1 is arranged by names of materials, subdivided by types of properties. These entries refer the user to a bibliography of some 13,000 items found in volume 2.

Noise

King, Richard L., comp. **Airport Noise Pollution: A Bibliography of Its Effects on People and Property.** Metuchen, N.J.: Scarecrow, 1973. 380 pp.

This work includes over 2,000 items. There is a directory of organizations, plus author and subject indexes.

CHAPTER 4

Dictionaries

Although this chapter deals primarily with special dictionaries, it also mentions briefly a few little-known features of general dictionaries. How many people, for example, have examined the contents page of even a small dictionary and noticed the index and all the extra guides and lists at the back—such as guides to punctuation and lists of biographical names (and their dates and significance), places, and abbreviations? The large unabridged dictionaries, deluxe versions of the small desk-sized editions, have even more information of this sort. Yet many readers have never discovered it.

The special dictionaries devoted to technical subjects can be divided into two groups: English-language and foreign-language.

English-language technical dictionaries usually give very detailed definitions and are often well illustrated. They may cover a broad subject area, such as all of engineering, or may be restricted to one subject, such as astronautics. They are especially helpful to those new to a field.

Bilingual dictionaries give terms in both English and one foreign language. They are usually divided into two sections, with one section listing the term first in the foreign language, and the other listing it first in English. They, too, may cover a broad area or be restricted.

Multilingual (or **polyglot**) dictionaries typically list their terms in columns, in each of four or five languages. One section usually cites each English term with all its equivalents in the foreign languages covered. Definitions, if any, are almost always—of necessity—brief. Multilingual works tend to be restrictive, since otherwise they would be exceptionally bulky.

Both bilingual and multilingual dictionaries enable a person with only a smattering of a language to translate a short work well enough to get a fair idea of the main points of the writing.

Needless to say, any technical dictionary goes out of date as technologies change, but the more basic it is and the broader its coverage the longer it will last.

SELECTED EXAMPLES
English Language
General and Technical

Ballentyne, D. W. G., and D. R. Lovett, eds. **A Dictionary of Named Effects and Laws in Chemistry, Physics, and Mathematics.** 3rd ed. London: Chapman & Hall, 1970, 335 pp.

> The alphabetical listing by the person for whom the law or effect is named includes mathematical expressions where necessary for clarity.

Collocott, T., and A. Dobson, eds. **Chambers Dictionary of Science and Technology.** Rev. ed. Edinburgh: Chambers, 1974. 1,344 pp.

> This work has over 50,000 entries covering 100 areas of knowledge in both pure and applied science.

Crowley, Ellen T., and Robert C. Thomas, eds. **Acronyms and Initialisms Dictionary.** 4th ed. Detroit: Gale Research Company, 1973. 635 pp.

> This guide to alphabetical designations, contractions, initialisms, and similar condensed appellations has over 100,000 entries—an increase of 23,000 over the previous edition. There will be a new cumulative supplement each year, titled **New Acronyms and Initialisms.**

DeSola, Ralph, ed. **Abbreviations Dictionary.** 4th ed. New York: American Elsevier, 1974. 428 pp.

> This dictionary contains more than 130,000 definitions and

entries, including abbreviations, anonyms, acronyms, contractions, initials, and nicknames.

McGraw-Hill Dictionary of Scientific and Technical Terms. Edited by Daniel N. Lapedes. New York: McGraw-Hill, 1974. 1,634 pp.

This outstanding dictionary contains around 100,000 entries representing all disciplines of science and technology. It includes more than 3,000 illustrations and drawings.

Moser, Reta C., comp. **Space-Age Acronyms: Abbreviations and Designations.** 2d ed. Washington, D.C.: IFI/Plenum, 1969. 534 pp.

This book contains some 15,000 acronyms, primarily in aeronautics and astronautics but with some other fields of engineering and technology included. It has supplements for military and naval designations of equipment.

Aeronautics and Astronautics

Gentle, Ernest J., and Lawrence W. Reithmaier, eds. **Aviation and Space Dictionary.** 5th ed. Fallbrook, Calif.: Aero Publishers, Inc., 1974. 272 pp.

Aerospace topics as well as related fields, such as geophysics, nuclear engineering, and data processing, are presented.

Marks, Robert W., ed. **New Dictionary and Handbook of Aerospace: With Special Sections on the Moon and Lunar Flight.** New York: Praeger, 1969. 525 pp.

The definitions are of moderate length, some more than one page long.

Nayler, Joseph L., comp. **Dictionary of Astronautics.** New York: Hart Publishing Company, 1964. 316 pp.

The definitions are generally concise. The book covers many abbreviations and code names as well as regular scientific and technical terms. It is illustrated.

Chemical Engineering and Chemistry

Bennett, H., ed. **Concise Chemical and Technical Dictionary.** 3d ed. London: Arnold, 1975. 1,216 pp.

This work contains over 55,000 definitions, including a large number of trade names. It covers many fields related to chemistry and has several useful appendixes on chemical structure, indicators, and so forth. It has been considerably updated since the 1962 edition.

Condensed Chemical Dictionary. 8th ed. Edited by G. G. Hawley. New York: Van Nostrand Reinhold, 1971. 971 pp.

This book gives names, formulas, specific properties, sources, derivations, hazards, uses, and shipping regulations for 18,000 entries. There is also an index of manufacturers of products that have a trademark.

Dodd, A. E., comp. **Dictionary of Ceramics.** Totowa, N.J.: Littlefield, Adams, 1964. 327 pp.

The dictionary includes terms representing a broad sense of the word "ceramics," namely, pottery, glass, vitreous enamels, refractories, clay building materials, cement, concrete, electroceramics, and special ceramics.

Grant, Julius, ed. **Hackh's Chemical Dictionary.** 4th ed. New York: McGraw-Hill, 1969. 738 pp.

This work covers terms for several related fields such as physics, engineering, mineralogy, astrophysics, pharmacy, agriculture, biology, and medicine. It contains many chemical structure diagrams.

Snell, Foster D., and C. T. Snell, comps. **Dictionary of Commercial Chemicals.** 3d ed. New York: Van Nostrand Reinhold, 1962. 714 pp.

This dictionary gives substantial descriptions of the properties of, sources of, and uses for thousands of chemical substances, arranged under categories such as waxes, calcium compounds, and organic dyes. It has a detailed subject index.

Civil Engineering

Harris, Cyril M. **Dictionary of Architecture and Construction.** New York: McGraw-Hill, 1975. 553 pp.

Building materials and building trades (e.g., air conditioning and heating, water supply, illumination, and acoustics) are included. There are over 1,700 illustrations in the margins of the pages.

Vollmer, Ernst, comp. **Encyclopedia of Hydraulics, Soil, and Foundation Engineering.** New York: American Elsevier, 1967, 398 pp.

This book could more aptly be described as a dictionary, since no entry is more than a few lines long. It has a small number of drawings.

See also **Mining and Minerals**

Computers and Data Processing

Condensed Computer Encyclopedia. Edited by Philip B. Jordain. New York: McGraw-Hill, 1969. 605 pp.

The encyclopedia gives brief descriptions of technical terms and some slang expressions. It contains occasional charts and drawings and is useful to many levels of readers.

Dictionary of Industrial Digital Computer Technology. Pittsburgh: Instrument Society of America, 1972. 120 pp.

Terms have been selected on the basis of their usage in industrial applications.

Maynard, Jeff. **Dictionary of Data Processing.** New York: Crane, Russak, 1976. 274 pp.

This dictionary defines more than 4,000 terms and includes terminology for hardware, interface methods, and software. Six appendixes include a list of common acronyms and abbreviations as well as an explanation of basic flowchart symbols.

Weik, Martin H., comp. **Standard Dictionary of Computers and Information Processing.** New York: Hayden, 1969. 326 pp.

The terms are defined at moderate length, and illustrations are occasionally used for clarity.

Electronics

Bones, R. A., ed. **Dictionary of Telecommunication.** New York: Philosophical Library, 1970. 200 pp.

Most of the definitions are brief but in a few cases cover several pages.

Graf, Rudolf F., comp. **Modern Dictionary of Electronics.** 4th ed. Indianapolis: Sams, 1972. 688 pp.

This work includes both theoretical terms as well as those pertaining to applications and technology. It includes a brief guide to pronounciation.

Markus, John, comp. **Electronics and Nucleonics Dictionary.** 3d ed. New York: McGraw-Hill, 1967. 743 pp.

> The dictionary gives definitions for over 16,000 terms; most are several lines long. There are frequent illustrations.

See also **Nuclear Engineering** and **Physics**

Mathematics

International Dictionary of Applied Mathematics. New York: Van Nostrand Reinhold, 1960. 1,173 pp.

> This work has definitions for the application of mathematics to 31 fields of physical science and engineering. Special tables list foreign terms in French, German, Russian, and Spanish and their English equivalents.

McDowell, C. H. **A Short Dictionary of Mathematics.** Totowa, N.J.: Littlefield, Adams, 1957. 103 pp.

> Part I deals with arithmetic and algebra; Part II concerns plane trigonometry and geometry. Definitions are concise, but mathematical expressions or drawings are used where needed for clarity.

Mechanical Engineering

Horner, J. G., comp. **Dictionary of Mechanical Engineering Terms.** 9th ed. Edited by G. K. Grahame White. New York: Heinman, 1967. 422 pp.

> Besides terms on mechanical power, this dictionary includes terms on metallurgy, metalworking, and other topics.

Nayler, J. L., and G. H. Nayler, **Dictionary of Mechanical Engineering.** New York: Hart Publishing Company, 1967. 406 pp.

> More or less restricted to mechanical engineering in the sense of the production and utilization of mechanical power, this work therefore excludes allied fields such as metallurgy and metrology. The definitions are concise.

Metals and Metallurgy

Osborne, A. K., comp. **An Encyclopedia of the Iron and Steel Industry.** 2d ed. New York: Heinman, 1967. 558 pp.

> This dictionary of terms, with definitions averaging several lines

in length, has a wide scope that includes metalworking, instruments, and so forth. It contains a detailed bibliography 28 pages long.

Mining and Minerals

Amstutz, G. C., comp. **Glossary of Mining Geology.** New York: American Elsevier, 1971. 196 pp.

This glossary defines terms involving mining and economic geology. There are cross-references to the equivalent terms in Spanish, French, and German.

Chambers's Mineralogical Dictionary. New York: Chemical Publishing, 1948. 87 pp.

This dictionary includes 40 pages of colored illustrations.

Nelson, A., and K. D. Nelson, eds. **Dictionary of Applied Geology: Mining and Engineering.** New York: Philosophical Library, 1967. 421 pp.

This book gives brief definitions of names of minerals, rock types, building materials, soils, and the like. It is intended primarily for those in mining and civil engineering.

U.S. Department of the Interior. **A Dictionary of Mining, Minerals, and Related Terms.** Compiled by Paul W. Thrush. Washington, D.C.: Superintendent of Documents, 1968. 1,269 pp.

This impressive compilation of about 55,000 terms contains definitions that are often quite detailed.

Nuclear Engineering

Hughes, L. E., and others, eds. **Dictionary of Electronics and Nucleonics.** New York: Barnes & Noble, 1970. 443 pp.

Besides concise definitions, this dictionary has over a hundred pages of supplementary matter such as properties of materials used in electronics, equivalency tables, electroacoustic data, and ionizing radiations.

See also **Electronics**

Physics

International Dictionary of Physics and Electronics. 2d ed. New York: Van Nostrand Reinhold, 1961. 1,355 pp.

This work consists mostly of short definitions a few lines long, but it has many cross-references. It includes mathematics and other fields of science and technology.

Thewlis, James, comp. **Concise Dictionary of Physics and Related Subjects.** New York: Pergamon, 1973. 366 pp.

This volume consists of short but adequate definitions covering not only physics but such related topics as astronomy, geophysics, mathematics, meteorology. It has many cross-references.

Bilingual Dictionaries

French

DeVries, Louis, and Stanley Hochman, comps. **French-English Science Dictionary.** 4th ed. New York: McGraw-Hill, 1975. 672 pp.

The dictionary includes a revised supplement of terms in such fields as aeronautics, electronics, atomic energy, and nuclear science. There is a "Grammatical Guide for Translators" section for those needing a review. Three thousand new terms have been added since the third edition.

Patterson, Austin M., comp. **French-English Dictionary for Chemists.** 2d ed. New York: Wiley, 1954. 476 pp.

The book defines over 42,000 terms, including many from several fields related to chemistry.

German

DeVries, Louis, comp. **Dictionary of Chemistry and Chemical Engineering.** New York: Academic, 1971, 1973. 2 vols.

Carefully chosen terms are used in this authoritative work. Volume 1 is German to English (1971) and volume 2 is English to German (1973).

DeVries, Louis, comp. **German-English Science Dictionary: For Students in Chemistry, Physics, Biology, Agriculture, and Related Sciences.** 3d ed. New York: McGraw-Hill, 1959. 592 pp.

Over 3,000 new terms have been added since the previous edition. The book contains a section entitled "Suggestions for Translators" to aid the inexperienced reader of German.

DeVries, Louis, and W. E. Clason, comps. **Dictionary of Pure and Applied Physics.** New York: American Elsevier, 1964. 2 vols.

Volume 1 is German to English and volume 2 is English to German. They include many terms involving engineering as well as pure physics.

DeVries, Louis, and T. M. Herrmann, comps. **English-German Technical and Engineering Dictionary.** 2d ed. New York: McGraw-Hill, 1967. 1,154 pp.

This is a companion to the DeVries German-English version.

DeVries, Louis, and T. M. Herrmann, **German-English Technical and Engineering Dictionary.** 2d ed. New York: McGraw-Hill, 1966. 1,178 pp.

This dictionary is a mate to the DeVries English-German work.

Dorian, A. F., comp. **Dictionary of Science and Technology (German-English and English-German).** New York: American Elsevier, 1967, 1970. 2 vols.

The combined length of the two volumes is over 2,000 pages. This is a set of significance.

Oppermann, Alfred, comp. **Dictionary of Modern Engineering.** 3d ed. New York: International Publications Services. 1972, 1973. 2 vols.

Volume 1 is English to German, and volume 2 is German to English. Together they form a useful compilation of engineering terms.

Patterson, A. M., comp. **German-English Dictionary for Chemists.** 3d ed. New York: Wiley, 1950. 541 pp.

This well-known work gives concise definitions for both pure and applied chemistry.

Schwenkhagen, H. F., comp. **Dictionary of Electrical Engineering: German-English, English-German.** New York: Wiley, 1963. 1,058 pp.

Besides the main focus, this book includes many items from related fields, such as mathematics, physics, chemistry, and metallurgy.

Italian

Denti, Renzo I., comp. **Italian-English, English-Italian Technical Dictionary.** 7th ed. New York: Heinman, 1970. 1,643 pp.

This work covers engineering as well as the pure sciences.

Russian

Alford, M. H. T., and V. L. Alford, comps. **Russian-English Scientific and Technical Dictionary.** Elmsford, N.Y.: Pergamon, 1970. 2 vols.

> Over 100,000 entries are given, covering 94 disciplines. Accent marks are included to aid pronounciation.

Callaham, Ludmilla I., comp. **Russian-English Chemical and Polytechnical Dictionary.** 3d ed. New York: Wiley, 1975. 896 pp.

> This dictionary can more accurately be described as covering most fields of engineering and the other technologies, thus serving as a general-purpose technical dictionary for this language. It now contains some 100,000 Russian terms, a 20 percent increase over the second edition.

Spanish

Sell, Lewis L., comp. **Comprehensive Technical Dictionary.** New York: McGraw-Hill, 1956, 1959. 2 vols.

> Volume 1 (in two parts) translates English to Spanish and volume 2 translates Spanish to English. They concern all areas of engineering and other technologies.

Multilingual Dictionaries

Broadbent, D. T., ed. **IFAC Multilingual Dictionary of Automatic Control Terminology.** Pittsburgh: Instrument Society of America, 1967. Variously paged.

> The terms are in English, French, German, Russian, Italian, and Spanish.

Burger, Erich, comp. **Technical Dictionary of Data Processing, Computers, and Office Machines.** Elmsford, N.Y.: Pergamon, 1970. 1,468 pp.

> The terms are given in English, German, French, and Russian.

Cagnacci-Schwicker, Angelo, comp. **International Dictionary of Metallurgy, Mineralogy, Geology, and the Mining and Oil Industries.** New York: McGraw-Hill, 1970. 1,530 pp.

> This multilingual approach (English, French, German, Italian) is arranged by English version, with separate indexes for the other languages, and has over 27,000 entries in each language.

DICTIONARIES 39

Clason, W. E., comp. **Dictionary of Chemical Engineering.** New York: American Elsevier, 1968. 2 vols.

> This dictionary gives terms in English, French, Spanish, Italian, Dutch, and German. Volume 1 deals with laboratory equipment, and volume 2 concerns processes and products.

Clason, W. E., comp. **Dictionary of Computers, Automatic Control and Data Processing.** 2d ed. New York: American Elsevier, 1971. 474 pp.

> This volume has nearly 4,000 entries. Terms are given in English, French, Spanish, Italian, Dutch, and German.

Clason, W. E., comp. **Dictionary of Electronics and Waveguides.** 2d ed. New York: American Elsevier, 1966. 833 pp.

> The foreign equivalents for English terms are given in French, Spanish, Italian, Dutch, and German. The definitions (in English) are brief. Separate Russian and Swedish supplements are available.

Clason, W. E., comp. **Dictionary of General Physics.** New York: American Elsevier, 1962. 859 pp.

> Arranged by English terms (with their definitions), this work gives foreign equivalents in French, Spanish, Italian, Dutch, and German.

Clason, W. E., comp. **Dictionary of Metallurgy.** New York: American Elsevier, 1967. 634 pp.

> The terms are in English (no definitions) with equivalents in French, Spanish, Italian, Dutch, and German. Each foreign language has an index of its own.

Clason, W. E., comp. **Dictionary of Nuclear Science and Technology.** 2d ed. New York: American Elsevier, 1970. 787 pp.

> The dictionary contains over 7,800 entries. Equivalent terms for the English are given in French, Spanish, Italian, Dutch and German.

Clason, W. E., comp. **Dictionary of Television, Radar, and Antennas.** New York: American Elsevier, 1955. 760 pp.

> This dictionary gives terms in English, French, Spanish, Italian, Dutch, and German. The definitions (in English) are brief.

Dorian, A. F., and J. Osenton, comps. **Dictionary of Aeronautics.** New York: American Elsevier, 1964. 842 pp.

> The languages are English, French, Spanish, Italian, Portuguese, and German. There are brief definitions (in English) as well as foreign equivalents for the English terms.

Goedecke, W., comp. **Dictionary of Electrical Engineering, Telecommunications, and Electronics.** New York, Ungar, 1964-1967. 3 vols.

> Terms are given in English, French, and German. Each language has a volume in which it is listed first.

International Academy of Astronautics. **Astronautical Multilingual Dictionary.** New York: American Elsevier, 1970. 936 pp.

> The foreign equivalents for the English terms are in Russian, German, French, Italian, Spanish, and Czech, with no definitions given.

James, Glenn, and R. C. James, comps. **Mathematics Dictionary.** 3d ed. New York: Van Nostrand Reinhold, 1968. 517 pp.

> The main section consists of English terms and their definitions. The separate indexes of terms in French, German, Russian, and Spanish, give the English equivalents.

Neidhardt, P., ed. **Technical Dictionary of Electronics.** Elmsford, N.Y.: Pergamon, 1967. 1,660 pp.

> The English, German, French, and Russian terms cover all aspects of electronics. Approximately 17,000 terms are defined.

Segditsas, P. E., comp. **Nautical Dictionary.** New York: American Elsevier, 1965-1966. 3 vols.

> The terms are in five languages: English, French, Italian, Spanish, and German. Volume 1 deals with general maritime terminology, volume 2 is on ships and their equipment, and volume 3 concerns marine engineering.

Sobecka, Z.; and others, eds. **Dictionary of Chemistry and Chemical Technology in Six Languages.** 2d ed. Elmsford, N.Y.: Pergamon, 1966. 1,325 pp.

> The dictionary consists of German, Spanish, French, Polish, and Russian equivalents of English terms. It has no definitions.

Sube, R., comp. **Dictionary of Nuclear Physics and Technology.** Elmsford, N.Y.: Pergamon, 1961. 1,606 pp.

> The languages are English, French, and Russian, with a separate section for each. Only equivalent terms are given; there are no definitions.

Visser, A., comp. **Telecommunication Dictionary.** New York: American Elsevier, 1960. 1,011 pp.

> The terms are in English, French, Spanish, Italian, German, and Dutch, with separate indexes for each language. No definitions are given.

Wyllie, R. J. M., and G. O. Argall, eds. **World Mining Glossary of Mining, Processing and Geological Terms.** San Francisco: Miller Freeman Pubs., 1975. 432 pp.

> The glossary contains around 11,000 terms in five languages (English, French, Swedish, German, and Spanish). It has an index for each language.

CHAPTER 5

Encyclopedias

In addition to the standard general-purpose encyclopedia, which answers most ordinary inquiries, engineers may need to consult the several technical encyclopedias that exist. These range from a few one-volume, rather skimpy works to at least one multivolume set that covers all of science and technology. There are also several multivolume works that are restricted to only one segment of a scientific discipline, such as applied chemistry.

Technical encyclopedias are **not** written for the expert, however, but for the technically inclined person who is just beginning to learn. An expert would probably find them too elementary, although some of the other sources of information they recommend might be of interest.

One good habit to cultivate in using an encyclopedia is that of checking the index volume for the entire set. Better encyclopedias have such a volume, and it can reveal entries worth seeing that might not otherwise come to mind. Not every article can include all the cross-references that a detailed index does.

SELECTED EXAMPLES

General—Science and Technology

Harper Encyclopedia of Science. Rev. ed. Edited by James R. Newman. New York: Harper & Row, 1967. 1,379 pp.

> The encyclopedia covers all areas of engineering and the other sciences with articles ranging from a few lines to several pages in

length. Photographs, drawings, and charts, many in color, illustrate it profusely. This is a useful quick source for a first look at a topic.

McGraw-Hill Encyclopedia of Science and Technology. 3d ed. New York: McGraw-Hill, 1971. 15 vols., supplements.

This very useful set (over 10,000 pages) deals with science and technology comprehensively. It is especially well edited and illustrated. Many of the signed articles are several pages long. There is a separate index volume. **McGraw-Hill Yearbook of Science and Technology** updates the encyclopedia annually. A separate readers' guide and a study guide are also available.

Van Nostrand's Scientific Encyclopedia. 4th ed. New York: Van Nostrand Reinhold, 1968. 2,008 pp.

All aspects of engineering and the other sciences are covered in around 16,500 articles, which range from a brief statement to several pages. The illustrations are well done, with many in color.

Astronautics

McGraw-Hill Encyclopedia of Space. New York: McGraw-Hill, 1968. 831 pp.

This encyclopedia covers rockets, satellites, space navigation, space flight, astrophysics and related topics. It is well-illustrated and includes many pages in color.

Biology

Gray, Peter, ed. **Encyclopedia of the Biological Sciences.** 2d ed. New York: Van Nostrand Reinhold, 1970. 1,027 pp.

All aspects of biology are included, excluding applied biological sciences. The work contains nearly 800 signed articles. There is also a subject index.

Chemical Engineering and Chemistry

Considine, D., ed. **Chemical and Process Technology Encyclopedia.** New York: McGraw-Hill, 1974. 1,261 pp.

This work coordinates data of interest to chemists, engineers, and managers. It emphasizes materials, equipment, and theory, covering topics related to chemistry, chemical engineering, and process engineering.

ENCYCLOPEDIAS 45

Hampel, Clifford A., ed. **Encyclopedia of the Chemical Elements.** New York: Van Nostrand Reinhold, 1968. 849 pp.

> The encyclopedia has signed articles (some over 10 pages long) on each element. There are also articles on general topics, such as electrode potentials and atomic abundance, that affect all elements.

Hampel, Clifford, and G. G. Hawley, eds. **Encyclopedia of Chemistry.** 3d ed. New York: Van Nostrand Reinhold, 1973. 1216 pp.

> This work discusses virtually every aspect of advances in chemistry since 1966. It contains over 500 articles, and any material retained from the previous edition has been updated.

Kingzett's Chemical Encyclopaedia: A Digest of Chemistry and Its Industrial Applications. 9th ed. Edited by D. H. Hey. New York: Van Nostrand Reinhold, 1967. 1,092 pp.

> This work covers both pure and applied chemistry. Some articles are several pages long. A detailed subject index supplements the main alphabetical arrangement.

Kirk, R. E., and D. F. Othmer, eds. **Encyclopedia of Chemical Technology.** 2d ed. New York: Wiley, 1963-   22 vols., plus supplements issued irregularly.

> This encyclopedia, with signed articles, contains a monumental and very thorough treatment of the subject. About half the articles deal with chemical substances, and some others concern industrial processes and fundamental topics.

Mead, William J., ed. **Encyclopedia of Chemical Process Equipment.** New York: Van Nostrand Reinhold, 1964. 1,065 pp.

> The articles, which are usually several pages long, cover classes of machinery and equipment, such as grinders, dryers, and steam generators. There are many illustrations.

Snell, Foster D., and others, eds. **Encyclopedia of Industrial Chemical Analysis.** New York: Wiley, 1966-1974. 20 vols.

> Alphabetically arranged, the articles are from a few lines to many pages long. The work is well illustrated, and has many cross-references.

Electronics

Susskind, Charles, ed. **Encyclopedia of Electronics.** New York: Van Nostrand Reinhold, 1962. 974 pp.

The encyclopedia has signed articles, often several pages long. Both theory and applications are discussed.

Environmental Science

McGraw-Hill Encyclopedia of Environmental Science. New York: McGraw-Hill, 1974. 700 pp.

This work covers a wide span of topics, such as meteorology, soils, mining, oceanography, civil engineering, ecology and conservation. Its 300 articles are well illustrated.

Instruments

Considine, Douglas M., comp. **Encyclopedia of Instrumentation and Control.** New York: McGraw-Hill, 1971. 788 pp.

The volume presents topics involved in the wide areas representing instruments and control systems, including their applications. An appendix lists a logical outline of the subjects covered.

Materials

Brady, George Stuart, ed. **Materials Handbook: An Encyclopedia for Purchasing Agents, Engineers, Executives, and Foremen.** 10th ed. New York: McGraw-Hill, 1971. 1,045 pp.

This book lists the basic properties and applications of thousands of metals, fabrics, chemical products, plastics, and so on. It is arranged alphabetically by general material names, with a specific subject index also.

Clauser, H. R., and others, eds. **Encyclopedia of Engineering Materials and Processes.** New York: Van Nostrand Reinhold, 1963. 787 pp.

The encyclopedia includes metals, nonmetallic materials, finishes and coatings, forms and shapes of materials, and the forming, casting, and fabricating processes for both metallics and nonmetallic materials. The articles average several pages in length and are signed.

Photography

Focal Encyclopedia of Photography. Rev. desk ed. New York: McGraw-Hill, 1969, 1,699 pp.

This very comprehensive book covers both theory and applications, with articles often several pages long. It is well illustrated.

Physics

Encyclopaedic Dictionary of Physics. Edited by J. Thewlis. Elmsford, N.Y.: Pergamon, 1961- . 9 vols., supplements issued irregularly.

This work has rather detailed signed articles on all aspects of physics, as well as some on mathematics, astronomy, physical metallurgy, physical chemistry, and so forth. Volume 8 contains subject and author indexes; volume 9 is a six-language dictionary. Supplements are issued every two years or so. Four were prepared by 1974.

Besancon, Robert M., ed. **Encyclopedia of Physics.** 2d ed. New York: Van Nostrand Reinhold, 1974. 848 pp.

The encyclopedia consists of 344 articles written by prominent authorities and includes related fields such as astrophysics and biophysics.

Plant Engineering

Woodley, Douglas R., ed. **Encyclopedia of Materials Handling.** Elmsford, N.Y.: Pergamon, 1964. 2 vols.

Profusely illustrated, the encyclopedia devotes sections to types of devices (conveyors, cranes, industrial trucks, and so on) as well as to processes and techniques (unitization, storage, loading and unloading, and the like).

Plastics

Encyclopedia of Polymer Science and Technology: Plastics, Resins, Rubbers, Fibers. Edited by Herman F. Mark and Norman G. Gaylord. New York: Wiley, 1964-1972. 16 vols.

This work has comprehensive signed articles, is well illustrated, and has many charts, graphs, and so on.

Modern Plastics Encyclopedia. New York: McGraw-Hill, 1941- [Annual]

This encyclopedia is issued to the subscribers to the periodical **Modern Plastics.** It describes the basic properties of many types of plastics as well as the processing and finishing of them. It includes many basic reference charts, a buyers' guide, and an index of trademarks.

Simonds, Herbert R., ed. **Encyclopedia of Plastics Equipment.** New York: Van Nostrand Reinhold, 1964. 599 pp.

This well illustrated encyclopedia describes around 200 types of processing equipment used in the plastics industry. The articles are signed, and are usually several pages long.

Simonds, Herbert R., and J. M. Church, eds. **Encyclopedia of Basic Materials for Plastics.** New York: Van Nostrand Reinhold, 1967. 500 pp.

The work has over 150 signed articles, usually several pages long. It contains many graphs, tables, and the like, and covers approximately 1,000 chemicals.

Safety

International Labour Organization. **Encyclopedia of Occupational Health and Safety.** New York: McGraw-Hill, 1972. 1,600 pp.

This work takes into account the most recent developments in industry, medicine, and accident prevention. Its concise articles cover a wide range of topics in this field, and it has ample cross-references.

Urban Planning

Whittick, Arnold, ed. **Encyclopedia of Urban Planning.** New York: McGraw-Hill, 1973. 1,200 pp.

This volume contains some 400 articles prepared by 70 authorities, cutting across several disciplines. It should be of interest to engineers, construction people, consultants, and those in fields such as ecology.

CHAPTER 6

Handbooks

A handbook is a summary of the highlights of a topic, written more as a review than as an instructional work. It is usually most useful to experts who want a quick review, and often lacks enough background or elementary material to be a good source for novices—although they might find it of use, at least for references.

Often a handbook is the product of many experts, who write chapters about their special fields. In most cases these chapters also include bibliographic references, which are usually well-known and important works.

Handbooks can now be found on a range of topics from all of physics to only semiconductors or pipes or noise control. Adequately detailed subject indexes are a must. Many of the best handbooks are updated every five years or so, but even if they are not they may remain useful for years beyond their publication dates. They are particularly good for tabular data and are extremely helpful for locating diverse, hard-to-find data, much of which does **not** become outdated quickly.

Some works with the term **handbook** in the title turn out to be subject dictionaries, with nothing more about a topic than a very brief description. Others concentrate on tabular data but do include a little textual material. Thus the borderline between handbooks, dictionaries, and tables of data is often vague.

SELECTED EXAMPLES

General—Engineering

CRC Handbook of Tables for Applied Engineering Science. Edited by Ray E. Bolz. Cleveland: CRC Press, 1973. 1,150 pp.

This book's 11 chapters deal with engineering materials, electrical engineering, chemical engineering, nuclear engineering, energy engineering, mechanical engineering, the human environment, safety, computation, measurements, etc. Written to provide the practicing engineer with a wide spectrum of data, it uses SI units as well as conventional units.

Kempe's Engineers Year-Book. 2 vols. West Wickham, Kent, Engl.: Morgan-Grampian, 1894-   [Annual]

This book provides an extremely broad coverage of most phases of engineering—civil, mechnical, metallurgical, aeronautical, and so forth—and also deals with materials, safety, legal matters, patents, and the like. It offers a vast amount of information. The edition for 1974 was its seventy-ninth.

Potter, James H., ed. **Handbook of the Engineering Sciences.** New York: Van Nostrand Reinhold, 1967. 2 vols.

Volume 1, **The Basic Sciences**, discusses the fundamental principles and formulas of mathematics, physics, chemistry, graphics, statistics, mechanics, and so on. Volume 2, **The Applied Sciences**, covers such areas as thermal phenomena, electronics, electromechanical energy conversion, astronautics, control systems, and materials science. The handbook is aimed at the first-year graduate level.

Souders, Mott, ed. **Handbook of Engineering Fundamentals.** 3d ed. New York: Wiley. 1975. 1,568 pp.

The aim of this work is to provide a wide range of engineers with basic information. Sections are devoted to such topics as mechanics, thermodynamics, electricity, and mathematics. Many tables and formulas are included.

Agricultural Engineering

Richey, C. B., and others, eds. **Agricultural Engineers' Handbook.** New York: McGraw-Hill, 1961. 880 pp.

The book has four sections, each containing chapters on the broad topics of crop-production equipment, soil and water conservation, farmstead structures and equipment, and basic agricultural data.

Air-Conditioning and Heating

American Society of Heating, Refrigerating, and Air-Conditioning Engineers. **ASHRAE Handbook and Product Directory.** New York: The Society, 1973-   3 vols.

> Each volume of this work is issued and updated independently but regularly. The **Equipment** volume describes the principles and operating features of typical models. The **Systems** volume concerns complete groups of equipment and their interrelationships. The **Applications** volume deals with the problems involved in various industries and environments. All volumes include manufacturers' literature.

American Society of Heating, Refrigerating, and Air-Conditioning Engineers. **ASHRAE Handbook of Fundamentals.** New York: The Society, 1972. 688 pp.

> The handbook presents both fundamental theories and basic design data. It also discusses materials and terminology and contains basic tables.

Strock, Clifford, and R. L. Koral, eds. **Handbook of Air-Conditioning, Heating and Ventilating.** 2d ed. New York: The Industrial Press, 1965. 1,472 pp.

> The 14 sections cover fundamental concepts such as load calculations and climatic data as well as the principles and equipment involved in heating, air-conditioning, refrigeration, piping, plumbing, and the like.

Astronautics

Goetzel, Claus G.; and others, eds. **Space Materials Handbook.** Reading, Mass.: Addison-Wesley, 1965. 624 pp.

> The four main parts of this book entitled "Space Environment," "Effect of Space Environment on Materials," "Materials in Space," and "Biological Interaction with Spacecraft Materials."

Haviland, R. P. **Handbook of Satellites and Space Vehicles.** New York: Van Nostrand Reinhold, 1965. 457 pp.

> Seventeen chapters include topics that range from the theory of orbits to materials in space and from space power to space environment.

Atomic Energy

Glasstone, Samuel. **Sourcebook on Atomic Energy.** 3d ed. New York: Van Nostrand Reinhold, 1967. 883 pp.

Chapters on the broad subject of atomic energy range from theoretical matters to nuclear reactors, applications of nuclear energy, and the like. The book has a minimum of mathematics.

U.S. Atomic Energy Commission. **Reactor Handbook.** 2d ed. New York: Wiley, 1960-1964. 5 vols.

The handbook consists of volume 1, **Materials** (1960); volume 2, **Fuel Reprocessing** (1961); volume 3, Part A, **Physics** (1962); volume 3, Part B, **Shielding** (1962); and volume 4, **Engineering** (1964). As the titles indicate, all phases of the subject are covered. The work stresses the materials and engineering topics more than the other aspects.

Chemical Engineering

Perry, John H., and Cecil H. Chilton, eds., **Chemical Engineers' Handbook.** 5th ed. New York: McGraw-Hill, 1973. 1,920 pp.

The more than two dozen chapters deal with fundamental topics (heat transfer, refrigeration, ion exchange, process control, and so forth). Several topics of related interest are also covered, such as electrical engineering, mechanical engineering, and cost estimation.

Civil Engineering

Merritt, Frederick S., ed. **Standard Handbook for Civil Engineers.** New York: McGraw-Hill, 1968. 1,326 pp.

This handbook includes the use of computers in civil engineering, specifications, construction management and materials, structural theory, foundation engineering, surveying, and municipal and regional planning as well as the engineering of highways, bridges, airports, rail transportation, tunnels, water supplies, sewage plants, harbors, and so on.

Seelye, Elwyn E., ed. **Data Book for Civil Engineers.** New York: Wiley, 1945-1960. 3 vols.

The work consists of volume 1, **Design** (3d ed., 1960); volume 2, **Specifications and Costs** (3d ed., 1957); and volume 3, **Field Practice** (2d ed., 1954). Volume 1 describes such topics as structural concrete, structural wood and plywood, and structural steel. Volume 2 discusses all types and applications of specifications and follows this with a study of cost factors in many kinds of situations. Volume 3 concerns the inspection and supervision of engineering construction projects.

Urquhart, Leon C., ed. **Civil Engineering Handbook.** 4th ed. New York: McGraw-Hill, 1959. 1,148 pp.

Ten sections deal with such subjects as surveying, mechanics of materials, hydraulics, stresses, soil mechanics, steel design, sewerage, water supply, and transportation engineering.

U.S. Department of the Interior. Bureau of Reclamation. **Design of Small Dams.** 2d ed. Washington, D.C.: Government Printing Office, 1973. 816 pp.

This handbook is aimed at serving as a guide for the design of small dams in the United States.

See also **Construction Engineering and Water Supply**

Computers and Data Processing

Grabbe, Eugene M.; and others, eds. **Handbook of Automation, Computation, and Control.** New York: McGraw-Hill, 1958-1961. 3 vols.

Volume 1, entitled **Control Fundamentals**, has 14 chapters on mathematics, followed by operations research and 8 chapters on feedback control. Volume 2, entitled **Computers and Data Processing**, covers the design and applications of both digital and analog computers. Volume 3, entitled **Systems and Components**, contains sections devoted to general considerations as well as to specific computer components and applications.

Harrison, Thomas J., ed. **Handbook of Industrial Control Computers.** New York: Wiley, 1972. 1,056 pp.

This work is devoted to the design, installation, and use of computer systems for industrial control. Topics include man/machine interface, system design factors, and application programming. There is a 27-page bibliography on computer applications.

Huskey, Harry D., and G. A. Korn, eds. **Computer Handbook.** New York: McGraw-Hill, 1962. 1,288 pp.

The book discusses the general design principles and the

utilization of both analog and digital computers, with equal emphasis on each computer type.

Klerer, Melvin, and G. A. Korn. **Digital Computer User's Handbook.** New York: McGraw-Hill, 1967. 750 pp.

> This handbook is intended for computer users other than those trained in programming or numerical analysis. Section 1 deals with programming, Section 2 with numerical techniques, Section 3 with statistical methods, and Section 4 with computer applications.

Sippl, Charles J., comp. **Computer Dictionary and Handbook.** Indianapolis: Sams, 1966. 766 pp.

> Nearly half the book contains brief definitions of terms. In addition there are over 20 supplements ranging from directories of computer service companies to lists of computer applications.

Construction Engineering

American Institute of Steel Construction. **Manual of Steel Construction.** 7th ed. New York: The Institute, 1970. Variously paged.

> Sections concern dimensions and properties of steel, beam and girder design, column design, connections, specifications and coding, and mathematical tables.

Burgess, R. A., and others, ed. **The Construction Industry Handbook.** 2d ed. Boston: Chambers, 1973. 467 pp.

> Sections are devoted to the financial aspects of construction, properties of building materials, and environmental design data (lighting, heating, acoustics, etc.).

Emerick, Robert H., ed. **Handbook of Mechanical Specifications for Buildings and Plants: A Checkbook for Engineers and Architects.** New York: McGraw-Hill, 1966. 482 pp.

> The handbook discusses specifications for power plants, heating systems, ventilating systems, and so forth. It gives a great amount of detailed information.

Gaylord, Edwin H., Jr., and C. N. Gaylord. **Structural Engineering Handbook.** New York: McGraw-Hill, 1968. 1,216 pp.

> The 26 chapters describe different types of structures

HANDBOOKS

(earthquake-resistant, timber, bridges) as well as more general topics (soil mechanics, computer use, erection methods, and so on).

Havers, John A., and F. W. Stubbs, Jr., eds. **Handbook of Heavy Construction.** 2d ed. New York: McGraw-Hill, 1971. 1,440 pp.

The work emphasizes three aspects of the subject: construction management, construction equipment, and construction applications. The latter includes steel construction, timber construction, piles and pile driving, river diversion, and tunneling.

LaLonde, William S., and M. F. Janes, eds. **Concrete Engineering Handbook.** New York: McGraw-Hill, 1961. 1,172 pp.

This work deals with various applications of concrete such as earthquake-resistant buildings, pavements, silos, and bridges as well as materials for reinforced concrete.

Merritt, Frederick S. **Building Construction Handbook.** 3d ed. New York: McGraw-Hill, 1975. 992 pp.

The book has sections on architecture, building materials, stresses, foundations, construction using different materials, heating and air conditioning, business aspects, etc.

Merritt, Frederick S., ed. **Structural Steel Designers' Handbook.** New York: McGraw-Hill, 1972. 886 pp.

This work concentrates on design criteria and structural analysis for both routine and complex structures. Chapters are devoted to topics such as properties of structural steels, design of building members, criteria for bridge design, as well as specific types of bridges.

O'Brien, James J., ed. **Construction Inspection Handbook.** New York: Van Nostrand Reinhold, 1974. 510 pp.

Written for inspectors of construction projects, this volume has chapters that center on the role of the inspector, features of construction (concrete, carpentry, finishes, etc.), and project management, including scheduling.

Waddell, Joseph J., ed. **Concrete Construction Handbook.** 2d ed. New York: McGraw-Hill, 1974. 960 pp.

The 14 chapters deal with topics such as properties of concrete, formwork and shoring, placing of concrete, precast and prestressed concrete, and repair of concrete. They also include toughness and creep of concrete, use of plastics in formwork, and modern surveying techniques.

Winterkorn, Hans F.; Fang, Hsai-Yang, eds. **Foundation Engineering Handbook.** New York: Van Nostrand Reinhold, 1975. 751 pp.

This book has 25 chapters on soil mechanics, drainage, soil stabilization, retaining walls, foundation vibration, pile foundations, etc.

Controls and Instruments

Considine, Douglas M., and S. D. Ross, eds. **Handbook of Applied Instrumentation.** New York: McGraw-Hill, 1964. 1,156 pp.

The handbook gives hundreds of typical applications of instruments in manufacturing, aviation, utilities, laboratories, and so forth. It evaluates different ways of measuring variables subject to instrumentation.

Kallen, Howard P., ed. **Handbook of Instrumentation and Controls.** New York: McGraw-Hill, 1961. 750 pp.

The discussion of measurements of basic phenomena such as temperatures and pressures is followed by more specific applications, including control devices. Applications cover boilers, steam turbines, diesel engines, and air-conditioning and heating systems.

See also **Computers and Data Processing** and **Electronics**.

Electrical Engineering

Berger, Carl, ed. **Handbook of Fuel Cell Technology.** Englewood Cliffs, N.J.: Prentice-Hall, 1968. 607 pp.

The book is divided into four areas: theory, electrochemical techniques, fuel-cell systems, and the economics of fuel cells.

Fink, Donald G., and J. M. Carroll, eds. **Standard Handbook for Electrical Engineers.** 10th ed. New York: McGraw-Hill, 1968. 2,300 pp.

This very comprehensive handbook continues the series formerly edited by A. E. Knowlton. The 29 chapters discuss all aspects of the subject such as generators, power plants (including nuclear types), motors, electronic data processing, and power transmission.

National Fire Protection Association. **National Electrical Code.** 4th ed. Boston: The Association, 1975. 832 pp. (ANSI C 1-1975)

> This frequently revised code (itself an American National Standard) aims to protect persons and objects from the dangers that may arise from the use of electricity. It covers wiring methods, general equipment, special locations (hazardous, for example), tables, charts, and the like.

Say, Maurice G., ed. **Electrical Engineering Design Manual.** 3d ed. New York: Barnes & Noble, 1962. 318 pp.

> The chapters deal with such topics as winding coils, chokes, and motors.

See also **Electronics.**

Electronics

EEE Editors. **Electronic Circuit Design Handbook.** 4th ed., rev. Blue Ridge Summit, Pa.: TAB, 1971. 410 pp.

> This collection of over 600 circuits (selected for their usefulness) was originally published in the **EEE** magazine. The circuits are arranged in 19 broad categories, such as pulse, protection, and amplifier.

Fink, Donald G., ed. **Electronics Engineers' Handbook.** New York, McGraw-Hill, 1975. 2,000 pp.

> There are 27 sections devoted to topics such as components, circuits, amplifiers, antennas, broadcasting, data processing.

Gruenberg, Elliot L., ed. **Handbook of Telemetry and Remote Control.** New York: McGraw-Hill, 1967. 1,344 pp.

> The work covers applications (industrial, space-system, and so forth) as well as more fundamental topics (information sampling, data reduction, and so on).

Hamsher, Donald H., ed. **Communication System Engineering Handbook.** New York: McGraw-Hill, 1967. 600 pp.

> The 25 chapters contain not only the basic concepts (for instance, transmission and switching) but also specific applications such as closed-circuit television, power-line carrier systems, mobile-radio services, and cost studies.

Harper, Charles A., ed. **Handbook of Electronic Packaging.** New York: McGraw-Hill, 1969. 1,000 pp.

The editor covers in 15 chapters a wide range of fundamental topics such as soldering and welding techniques as well as some specific ones (packaging of microelectronics and hybrid systems, packaging for military applications, and so forth).

Harper, Charles A., ed. **Handbook of Materials and Processes for Electronics.** New York: McGraw-Hill, 1970. Variously paged.

The handbook has 15 chapters devoted to subjects such as wires and cables, ferrous metals, thin films, photofabrication, and elastomers.

Hughes, L. E. C., and F. W. Holland, eds. **Electronic Engineer's Reference Book.** 3d ed. London: Heywood Books, 1967. 1,532 pp.

This book discusses topics ranging from testing to computers and from automation to acoustics. It has very broad coverage.

Hunter, Lloyd P., ed. **Handbook of Semiconductor Electronics.** 3d ed. New York: McGraw-Hill, 1970. 1,100 pp.

Four main sections deal with the physics of materials and devices, techniques of materials and devices, circuit design, and measurement and analytical techniques.

Markus, John, ed. **Guidebook of Electronic Circuits.** New York: McGraw-Hill, 1974. 900 pp.

This work gives descriptions of around 3,000 circuits, using illustrated abstracts from over 3,600 articles published in recent years. There is a detailed index.

Markus, John, ed. **Sourcebook of Electronic Circuits.** New York: McGraw-Hill, 1968. 888 pp.

This comprehensive work gives diagrams and brief descriptions for over 3,000 circuits, arranged in more than 100 circuit categories such as noise, chopper, amplifier, and video. It has a detailed subject index.

Reference Data for Radio Engineers. 6th ed. Edited by H. P. Westman. Indianapolis: Sams, 1975. Variously paged.

Forty-five chapters include such topics as basic units, design of filters, transistor circuits, properties of materials, nuclear physics, and mathematical tables. The book is well illustrated and concentrates on formulas and tabular data.

HANDBOOKS

Sucher, Max, and J. Fox, eds. **Handbook of Microwave Measurements.** 3d ed. New York: Wiley, 1963. 3 vols.

> Volume 1 concerns some fundamentals aspects such as measurement of power and attenuation. Volume 2 includes measurement of Q, dielectric constant, breakdown voltage, and so forth. Volume 3 has chapters on receiver characteristics, antenna measurement, transmission line charts, and the like.

Thomas, Harry E., ed. **Handbook of Integrated Circuits.** Englewood Cliffs, N.J.: Prentice-Hall, 1973. 346 pp.

> This work discusses the principles of solid-state electronics as well as the techniques of constructing microminiature devices. It has many useful appendixes, such as a glossary, descriptions of special tools and test devices, etc.

See also **Electrical Engineering** and **Radar**

Electroplating

Graham, Arthur Kenneth, ed. **Electroplating Engineering Handbook.** 3d ed. New York: Van Nostrand Reinhold, 1971. 845 pp.

> Part 1 is confined to fundamentals, such as water requirements and analysis of plating baths. In Part 2 more detailed topics are discussed, for instance, auxiliary equipment for the plating room, barrels, and racks.

Environmental Engineering

CRC Handbook of Environmental Control. Edited by Richard G. Bond and Conrad P. Straub. Cleveland: CRC Press, 1973-1974. 4 vols.

> Volume 1 covers air pollution, volume 2 covers solid waste, volume 3 covers water supply and treatment, and volume 4 is on waste-water treatment and disposal. The emphasis is on tabular data. Both scientific and socioeconomic factors are included.

Zilly, Robert G. ed. **Handbook of Environmental Civil Engineering.** New York: Van Nostrand Reinhold, 1975. 1,042 pp.

> This volume emphasizes the role of environmental controls as they affect soil and foundation engineering, sanitary engineering, facilities for mass transportation, etc.

See also **Pollution**

Fluids

See **Hydraulics**

Gas Engineering

Gas Engineers Handbook: Fuel Gas Engineering Practices. New York: The Industrial Press, 1965. Variously paged.

> Sponsored by the American Gas Association, the handbook consists of 15 sections (each with several chapters), including subjects such as properties of gases, production of gas, testing, storage, distribution, and utilization.

Katz, Donald L., and others, eds. **Handbook of Natural Gas Engineering.** New York: McGraw-Hill, 1959. 802 pp.

> This volume treats natural gas in its various stages, from occurrence in the reservoirs through production, processing, and transportation. Major emphasis is given to the properties of the hydrocarbons and the application of these properties in flow and processing operations.

Heat Transfer

Rohsenow, W. M., and J. Hartnett, eds. **Handbook of Heat Transfer.** New York: McGraw-Hill, 1972. 1,518 pp.

> This work has 19 chapters devoted to such topics as basic principles, boiling, radiation, thermodynamic tables, special applications, heat exchangers, etc.

Heating

See **Air-Conditioning and Heating**

Highway Engineering

Baker, Robert F., ed. **Handbook of Highway Engineering.** New York: Van Nostrand Reinhold, 1975. 768 pp.

> This handbook covers a broad range of topics affecting all aspects of highway engineering, such as planning and economics, traffic control and intersection design, drainage and foundations, tunnels and ramps, and maintenance and snow control.

Hydraulics

Creager, William P., and J. D. Justin, eds. **Hydroelectric Handbook.** 2d ed. New York: Wiley, 1950. 1,151 pp.

> General topics such as rainfall and evaporation, dams, conduits, powerhouses, and equipment constitute the main sections of this book, which is still useful for many purposes.

Davis, Calvin V., and K. E. Sorensen, eds. **Handbook of Applied Hydraulics.** 3d ed. New York: McGraw-Hill, 1969. 1,216 pp.

>The contents include canals, dams, hydraulic machinery, flood control, irrigation, water supplies, sewage treatment, and groundwater.

Streeter, Victor L., ed. **Handbook of Fluid Dynamics.** New York: McGraw-Hill, 1961. 1,228 pp.

>The subject matter ranges from basic equations for different types of flow to specific equipment such as fluid power transmission to magnetohydrodynamic topics.

Hydrology

Chow, Ven-Te, ed. **Handbook of Applied Hydrology: A Compendium of Water-Resources Technology.** New York: McGraw-Hill, 1964. 1,468 pp.

>The handbook is divided into four groupings. The first discusses closely related sciences (oceanography, meteorology, geomorphology, and the like). The second covers the hydrologic cycle (rainfall, groundwater, runoff, river sedimentation, and so on). The third concerns applications of hydrology (flood routing, river forecasting, urban hydrology, and so forth). The fourth is on the socioeconomic aspects (water laws, water policy, and the like).

See also **Water Supply**

Industrial Engineering

Hartman, W.; and others, eds. **Management Information Systems Handbook.** 2d ed. New York: McGraw-Hill, 1968. Variously paged.

>The book's five parts introduce systems engineering, project management, feasibility studies, systems development, and system implementation and evaluation.

Ireson, William Grant, and E. L. Grant, eds. **Handbook of Industrial Engineering and Management.** 2d ed. Englewood Cliffs, N.J.: Prentice-Hall, 1971. 907 pp.

>This book is aimed not only at industrial engineers but also at engineers now engaged in management activities. Chapters range from managerial economics to critical path methods, from industrial safety to attitudes of labor towards industrial engineering methods, and so on. The handbook includes bibliographies.

Machol, Robert E., and others, eds. **System Engineering Handbook.** New York: McGraw-Hill, 1965. 1,084 pp.

> Prepared for those designing large, complex systems, this work describes semiautomatic devices, man-computer interaction, and other elements and components.

Maynard, H. B., ed. **Industrial Engineering Handbook.** 3d ed. New York: McGraw-Hill, 1971. 1,984 pp.

> This new edition is considerably updated with new material also added. This authoritative work deals with a wide spectrum of material of interest to the industrial engineer.

See also **Plant Engineering and Manufacturing**

Instruments

See **Controls and Instruments**

Lighting

Illuminating Engineering Society. **IES Lighting Handbook: The Standard Lighting Guide.** 5th ed. New York: The Society, 1972. Variously paged.

> Many applications of lighting (educational, residential, industrial, transportational, and so forth) follow a series of chapters on the basics of color, vision, measurements, and related topics.

Lubrication

O'Connor, James J.; and John Boyd, eds. **Standard Handbook of Lubrication Engineering.** New York: McGraw-Hill, 1968. 1,000 pp.

> The American Society of Lubrication Engineers sponsors this review of standard practices. Part 1 is on principles; Part 2 covers general practices. Part 3 deals with special equipment, and Part 4 discusses problems of specific industries, including metal, textile, railroad, and nuclear plant.

Manufacturing

See **Plant Engineering and Manufacturing**

Materials

CRC Handbook of Materials Science. Edited by Charles T. Lynch. Cleveland: CRC Press, 1974. 500 pp.

HANDBOOKS

This work, largely tabular in content, provides a guide to data on properties of metals, oxides, glasses, polymers, composites, electronic materials, nuclear materials, biomedical materials, etc.

Electronic Properties Information Center. **Handbook of Electronic Materials.** New York: Plenum, 1971. 3 vols.

Volume 1 concerns optical materials, and presents properties and characteristics. Volume 2 covers semiconducting compounds. Volume 3 covers silicon nitride for microelectronic applications. The work cites over 1,100 references.

Lowenheim, Frederick A., and Marguerite K. Moran. **Faith, Keyes & Clark's Industrial Chemicals.** 4th ed. New York: Wiley, 1975. 904 pp.

This is an updated version of a well-known title. It thoroughly describes 140 products, such as glycerine, acrylates, cyclohexane, and so on. It covers production techniques, use patterns, economic history, current U.S. manufacturers, and similar data.

Moss, J. B., ed. **Properties of Engineering Materials.** Cleveland: CRC Press, 1971. 376 pp.

The book gives descriptions of the properties of materials now available as well as an analysis of future developments. It also has a summary on failure of materials.

Parker, Earl R., ed. **Materials Data Book for Engineers and Scientists.** New York: McGraw-Hill, 1967. 398 pp.

Sixteen chapters describe the properties of such materials as various types of alloys, ceramics, concrete, and plastics as well as give information about corrosion and welding techniques.

Samsonov, G. V., ed. **The Oxide Handbook.** New York: IFI/Plenum, 1973. 524 pp.

This translation and updating of the Russian of 1969 presents tables on the following properties: thermal, mechanical, electrical, magnetic, optical, nuclear, chemical, and refractory. It has over 750 references.

Shand, E. B., ed. **Glass Engineering Handbook.** 2d ed. New York: McGraw-Hill, 1958. 484 pp.

The handbook's four divisions, containing several chapters each, deal with glass technology, glass manufacture, applications of glass, and fibrous glass.

See also **Construction Engineering, Metals and Metal Working,** and **Plastics**

Mathematics

Bronshtein, Ilia N., and K. A. Semendyayev, comps. **Guide Book to Mathematics for Technologists and Engineers.** Elmsford, N.Y.: Pergamon, 1964. 783 pp.

> Besides a large quantity of tables, the book has descriptive matter that discusses theorems and principles. Section headings are: "Tables and Graphs," "Elementary Mathematics," "Analytic and Differential Geometry," and "Analysis and Interpretation of Experimental Results."

Grazda, Edward E., ed. **Handbook of Applied Mathematics.** 4th ed. New York: Van Nostrand Reinhold, 1966. 1,119 pp.

> This handbook describes the important principles of mathematics (ranging from arithmetic through calculus), and the applications of mathematics to different trades and technologies, such as brickworking, heating, electronics, carpentry, and so forth. It includes many problems and examples.

Hicks, Tyler G., ed. **Standard Handbook of Engineering Calculations.** New York: McGraw-Hill, 1972. 1,206 pp.

> This volume presents step-by-step procedures for making more than 2,000 calculations involving almost all routine problems found in daily engineering practice. It has a separate chapter for each of twelve basic types of engineering, such as civil, mechanical, sanitary, chemical, etc.

Korn, Granino, and T. M. Korn, eds. **Mathematical Handbook for Scientists and Engineers.** 2d ed. New York: McGraw-Hill, 1968. 1,130 pp.

> This comprehensive collection of definitions, theorems, and formulas plus some numerical tables (ranging from squares to Chebyshev polynomials) includes both undergraduate and graduate subjects.

Pearson, Carl E., ed. **Handbook of Applied Mathematics: Selected Results and Methods.** New York: Van Nostrand Reinhold, 1974. 1,265 pp.

> The 21 chapters are devoted to such topics as optimization techniques, transform methods, wave propagation, tensors, and

partial differential equations. Each technique or formula is carefully explained.

Tuma, Jan J., ed. **Handbook of Physical Calculations.** New York: McGraw-Hill, 1976. 384 pp.

This work, written for practicing engineers, presents a summary of the major definitions, formulas, tables, and examples of technological physics. It can also be used as a dictionary. Chapters cover statics and dynamics of rigid bodies, mechanics of deformable solids, fluid mechanics, mechanics of heat and gases, electricity, vibrations, acoustics, etc.

Mechanical Engineering

Baumeister, T., and L. Marks, eds. **Standard Handbook for Mechanical Engineers.** 7th ed. New York: McGraw-Hill, 1967. 2,464 pp.

This comprehensive work encompassing materials, mechanical properties and phenomena, mathematics, power sources, transportation, electronics, mechanisms, optics, and dozens of other subjects is extremely useful.

Flugge, Wilhelm, ed. **Handbook of Engineering Mechanics.** New York: McGraw-Hill, 1962. 1,632 pp.

This handbook divides the subject into seven areas: mathematics, mechanics of rigid bodies, theory of structures, elasticity, plasticity and viscoelasticity, vibrations, and fluid mechanics.

Jones, Franklin D., and others, eds. **Ingenious Mechanisms for Designers and Inventors.** New York: The Industrial Press, 1930-1967. 4 vols.

This work describes the principles and operations of hundreds of mechanisms. Each is carefully described, and many are illustrated. Typical chapter headings are: "Cams," "Reciprocating Mechanisms," and "Reversing Mechanisms."

Kent, R. T. **Mechanical Engineers' Handbook.** 12th ed. New York: Wiley, 1950. 2 vols.

Part 1, **Design and Production**, has chapters on materials, motors, controls, working of metals, materials handling, and so forth. Part 2, **Power**, concerns turbines, internal combustion engines, transportation, heating, electric power, instrumentation, and so on.

Rothbart, Harold P., ed. **Mechanical Design and Systems Handbook.** New York: McGraw-Hill, 1964. 1,594 pp.

> Typical chapter topics are: "Dynamics of Materials," "Power Devices," and "Systems Analysis."

Metals and Metalworking

American Society for Metals. **Metals Handbook.** 8th ed. Metals Park, Ohio: The Society, 1961-1976. 10 vols.

> Topics covered include machining, forming, selection of metals, forging and casting, welding and metallography, fractography, and failure analysis and prevention.

American Society of Mechanical Engineers. **ASME Handbook: Metals Engineering—Design.** 2d ed. Edited by Oscar J. Horger. New York: McGraw-Hill, 1965. 619 pp.

> The book's five sections are devoted to selection of materials, mechanical properties of metals in design, other physical properties affecting design, nondestructive testing, and design considerations.

American Society of Mechanical Engineers. **ASME Handbook: Metals Engineering—Processes.** Edited by Roger W. Bolz. New York: McGraw-Hill, 1958. 428 pp.

> The chapters are grouped into the following broad categories: heat treatment of steel, casting, hot working, cold working, powder metallurgy, welding and cutting, machining, finishing, and electroforming.

American Society of Mechanical Engineers. **ASME Handbook: Metals Properties.** Edited by Samuel S. Hoyt. New York: McGraw-Hill, 1954. 433 pp.

> For each type of metal, the handbook gives physical properties, mechanical properties, chemical composition, critical points, and so forth. It has an index arranged by chemical composition.

American Society of Tool and Manufacturing Engineers. **ASTME Die Design Handbook.** New York: McGraw-Hill, 1965. 788 pp.

> The book describes hundreds of practical designs from all types of dies used in cold pressworking. It also characterizes presses, accessories, and materials.

HANDBOOKS

American Society of Tool and Manufacturing Engineers. **Handbook of Fixture Design.** New York: McGraw-Hill, 1962. 479 pp.

This handbook provides hundreds of actual fixture designs for all types of metalworking operations and includes both predesign analysis and design procedures.

American Society of Tool and Manufacturing Engineers. **Tool Engineers Handbook.** 3d ed. New York: McGraw-Hill, 1976. 2,400 pp.

This comprehensive work has over 100 chapters dealing with a wide range of topics (grinding, brazing, quality control, cams, fluid power, plant layout, and so on). It has been extensively updated since the last edition. An added section on the metric system should be helpful.

American Welding Society. **Welding Handbook.** 6th ed. New York: The Society. 1971. 5 vols.

The volumes deal with fundamentals of welding, welding processes, special welding and cutting processes, metals and their weldability, and applications.

Bunshah, R. F., and Gray, R., eds. **Techniques of Metals Research.** New York: Wiley, 1968-

This series of reference books will provide a comprehensive account of all aspects of metals research. To date it has produced volumes on metals preparation, observation of structure, measurement of mechanical properties, analytical techniques, and metallographic techniques. R. Gray edited only the metallographic techniques volume.

Hampel, Clifford A., ed. **Rare Metals Handbook.** 2d ed. New York: Van Nostrand Reinhold, 1961. 715 pp.

Some 30 chapters are devoted to one metal each, such as cobalt, indium, and rhenium, and give its properties, applications, production methods, and other features. A few chapters discuss generalities of a metallurgical nature.

Leslie, W. H. P., ed. **Numerical Control Users' Handbook.** New York: McGraw-Hill, 1970. 482 pp.

This collection of chapters covers the operational, economic, and technical aspects of this application of automation to metalworking. The work includes a useful glossary.

Machinery's Handbook: A Reference Book for the Mechanical Engineer, Draftsman, Toolmaker, and Machinist. 20th ed., rev. Edited by Erik Oberg and F. D. Jones. New York: The Industrial Press, 1975. 2,482 pp.

> This compilation of tables deals with materials, threads, machining, jigs, metalworking, metal properties, and so forth. There is textual description of major points.

Ross, Robert B., ed. **Metallic Materials.** London: Chapman & Hall, 1968. 936 pp.

> This book lists the composition and properties of some 25,000 alloys. It has an index of trade names and also a directory of manufacturers.

Van Horn, Kent, ed. **Aluminum.** Metals Park, Ohio: American Society for Metals, 1966. 3 vols.

> Volume 1 deals with properties, physical metallurgy, and phase diagrams; volume 2 is on design and applications; volume 3 concerns fabrication and finishing. This work is a thorough treatment of the subject.

Woldman, Norman E. **Engineering Alloys.** 5th ed. New York: Van Nostrand Reinhold, 1973. 1,440 pp.

> This book has two main sections. The first lists alloys by serial number, giving trade name, composition, properties, and uses. The second is arranged alphabetically by trade names and corresponding serial numbers. Indexes by manufacturer and certain tabular data are included. There have been 15,000 new alloys added since the previous edition.

Mining and Minerals

Hurlbut, Cornelius S., Jr. **Dana's Manual of Mineralogy.** 18th ed. New York: Wiley, 1971. 579 pp.

> This well-known reference source serves as a sort of handbook. Most of the book describes the properties, occurrences, and uses of around 200 minerals. The other portions of the work discuss crystallography, physical mineralogy, chemical mineralogy, and the like. There is a subject and mineral index.

Peele, R., ed. **Mining Engineers' Handbook.** 3d ed. New York: Wiley, 1941. 2 vols.

HANDBOOKS

The contents range from tunneling to mine ventilation to ore sampling. A useful source of information in spite of its age.

SME Mining Engineering Handbook. Edited by Arthur B. Cummins and Ivan A. Given. New York: Society of Mining Engineers of AIME, 1973. 2 vols.

These volumes contain 35 chapters which deal with all aspects of mining, ranging from safety to rock mechanics, from strip mining to mineral processing a valuable work for this field.

Noise

Harris, Cyril M., ed. **Handbook of Noise Control.** New York: McGraw-Hill, 1957. 1,184 pp.

The handbook has 40 chapters devoted to a wide variety of topics, both fundamental, such as sound propagation and hearing mechanism, and applied, such as noise in rail transportation, gears, and air conditioners.

Nuclear Engineering

See **Atomic Energy**

Ocean Engineering and Oceanography

Myers, John J.; and others, eds. **Handbook of Ocean and Underwater Engineering.** New York: McGraw-Hill, 1969. 800 pp.

These 12 chapters are on basic oceanography, underwater cables, fixed structures, diving, wind and wave loads, ocean operations, and so forth.

Sverdrup, H. U.; and others. **The Oceans: Their Physics, Chemistry, and General Biology.** Englewood Cliffs, N.J.; Prentice-Hall, 1942. 1,087 pp.

This book is still a well-known and heavily used source. It devotes 20 chapters to all aspects of oceanography, ranging from a study of tides to marine sedimentation.

Petroleum

Bell, Harold S., ed. **Petroleum Transportation Handbook.** New York: McGraw-Hill, 1963. 510 pp.

Besides many chapters on pipelines, topics include rail, vehicular, and marine transportation of petroleum in its many forms.

Bland, William F., and R. L. Davidson, eds. **Petroleum Processing Handbook.** New York: McGraw-Hill, 1967. 1,102 pp.

> Fourteen sections cover processes, equipment, materials of construction, chemicals, safety, instruments, and other related subjects. One section is devoted to sources of information.

Photography

Thomas, Woodlief, Jr., ed. **SPSE Handbook of Photographic Science and Engineering.** New York: Wiley, 1973. 1,416 pp.

> A publication of the Society of Photographic Scientists and Engineers, this far-reaching book has chapters on radiation sources, photographic optics, filters, color photography, densitometry, image structure, projection and viewing, microphotography test methods, etc.

Physics

American Institute of Physics. **Physics Handbook.** Edited by Dwight E. Gray. 3d ed. New York: McGraw-Hill, 1972. 2,200 pp.

> This handbook treats in detail large, basic topics such as heat, nuclear physics, optics, and electricity and magnetism. It has many charts and tables.

Condon, E. U., and H. Odishaw, eds. **Handbook of Physics.** 2d ed. New York: McGraw-Hill, 1967. 1,690 pp.

> This book contains sections, of several chapters each, on mathematics, mechanics, electricity and magnetism, heat and thermodynamics, optics, and atomic, solid-state, and nuclear physics.

Menzel, Donald H., ed. **Fundamental Formulas of Physics.** New York: Dover, 1960. 2 vols.

> Organized into chapters such as basic mathematical formulas, statistics, nomograms, physical constants, mechanics, relativity, heat and thermodynamics, electronics, cosmic rays, particle accelerators, meteorology, and biophysics, the work lists and discusses important formulas. It has a subject and name index.

Piping

Crocker, Sabin, and Reno C. King, eds. **Piping Handbook.** 5th ed. New York: McGraw-Hill, 1967. 1,652 pp.

> In this book many special types of piping are referred to (fire-

protection, underground-steam, water-supply, hydraulic-power) as well as more general topics such as fluid mechanics, insulation, and corrosion.

Plant Engineering and Manufacturing

American Society of Tool and Manufacturing Engineers. **Manufacturing Planning and Estimating Handbook.** New York: McGraw-Hill, 1963. 849 pp.

> The handbook discusses manufacturing analysis, materials-handling analysis, job ratings, inspection and quality control, final manufacturing estimates, and so on.

Cagle, Charles V., comp. **Handbook of Adhesive Bonding.** New York: McGraw-Hill, 1973. 800 pp.

> Based on the work of twenty-seven experts, the book describes bonding methods in a variety of applications.

Davidson, A., ed. **Handbook of Precision Engineering.** New York: McGraw-Hill, 1970-

> Each volume will deal with one aspect of the major subject. Volume 1 concerns fundamentals and includes design theory, kinematics, quality control, and the like. Volume 2 is on materials. Other volumes cover such topics as joining techniques, machining processes, and precision measurement. By 1976 ten volumes had been published.

Greene, J. H. **Production and Inventory Control Handbook.** New York: McGraw-Hill, 1970. 800 pp.

> This book covers practices and procedures in such areas as forecasting, manpower requirements, manual and computer processing, model stimulation, and linear programming.

Heyel, Carl, ed. **The Foreman's Handbook.** 4th ed. New York: McGraw-Hill, 1967. 591 pp.

> Its treatment of the subject would make this volume useful to industrial engineers and plant managers as a means of understanding the role of the foreman.

Lewis, Bernard T., and J. P. Marron, eds. **Facilities and Plant Engineering Handbook.** New York: McGraw-Hill, 1974. 1,024 pp.

> Written for those responsible for the facilities required to support production processes, such as facilities engineers, maintenance

managers, and plant engineers, these chapters treat a variety of topics ranging from plant layout to refrigeration equipment, from pollution control to budgets.

Morrow, L. C., ed. **Maintenance Engineering Handbook.** 2d ed. New York: McGraw-Hill, 1966. Variously paged.

Aimed at the plant engineer or those responsible for the upkeep of plant equipment and services, this book has separate sections on the maintenance of different types of equipment (mechanical, electrical), lubricants, welding, corrosion, schedules and costs, and so forth.

Production Handbook. Edited by Gordon B. Carson. 3d ed. New York: Ronald, 1972. Variously paged.

The handbook covers a wide scope of subjects related to production (purchasing, work simplification, operations research, plant layout, plant maintenance, and so on).

Staniar, William, ed. **Plant Engineering Handbook.** 2d ed. New York: McGraw-Hill, 1959. Variously paged.

This book has chapters on a broad range of topics related to the technical side of plant operation such as motors, lubrication, vibration, welding, heating of industrial buildings, waste disposal, and plastics and rubber.

Plastics

Harper, Charles, ed. **Handbook of Plastics and Elastomers.** New York: McGraw-Hill, 1975. 950 pp.

This work provides data, performance application information, and guidelines for the use of the entire range of plastics and elastomers.

Lee, Henry, and Kris Neville, eds. **Handbook of Epoxy Resins.** New York: McGraw-Hill, 1966. 960 pp.

This thorough review of the subject describes thousands of industrial applications.

Lubin, George, ed. **Handbook of Fiberglass and Advanced Plastics Composites.** New York: Van Nostrand Reinhold, 1969. 894 pp.

This handbook offers a wide coverage of fiber glass and high modulus composites, ranging from filaments and high temperature resins through bag molding and machining of plastics. Design and testing techniques are also included.

Mohr, J. Gilbert, and Samuel F. Oleesky, eds. **SPI Handbook of Technology and Engineering of Reinforced Plastics/Composites.** 2d ed. New York: Van Nostrand Reinhold, 1973. 416 pp.

Designed to be used in calculating configurations for reinforced plastics products, this work covers molding methods, tool and product design, and properties of materials, as well as test methods, safety, and future outlook for growth in this field.

Society of the Plastics Industry, Inc. **Plastics Engineering Handbook.** 3d ed. New York: Van Nostrand Reinhold, 1960. 565 pp.

This work describes the processes, finishing, and application of all types of plastics as well as performance tests, standards, and so forth.

Pollution

Cheremisinoff, P. E., and Young, Richard A., eds. **Pollution Engineering Practice Handbook.** Ann Arbor: Ann Arbor Science, 1975. 500(?) pp.

Topics covered include wet scrubbers, stack design, filters, oil spills, solid wastes, noise measurement, etc. All aspects of pollution control are treated.

Ciaccio, Leonard L., ed. **Water and Water Pollution Handbook.** New York: Marcel Dekker, Inc., 1971. 4 vols.

Part 1 discusses environmental systems such as characteristics of water resources and waste effluents as well as effects of pollution. Part 2 deals with chemical, physical, bacterial, viral, instrumental, and bioassay techniques of analysis.

Lund, Herbert F., ed. **Industrial Pollution Control Handbook.** New York: McGraw-Hill, 1971. Variously paged.

The book is divided into three main sections. The first concerns basic background (history, standards, programs, and the like). The second part discusses the problems of each of 11 industrial areas (steel, paper, and so forth). The last section is devoted to control equipment and its operation.

Production Engineering

See **Plant Engineering and Manufacturing**

Quality Control and Reliability

Ireson, W. Grant, ed. **Reliability Handbook.** New York: McGraw-Hill, 1966. 702 pp.

> This handbook covers all angles of the subject, ranging from testing programs to the application of mathematics and statistics. Cost factors and organizational aspects are also included.

Juran, Joseph M., ed. **Quality Control Handbook.** 3d ed. New York: McGraw-Hill, 1974. 1,600 pp.

> The work discusses the economic aspects, specifications, organizational problems, statistical control methods, policies and objectives, and so on of quality control.

Kozlov, B. A., and I. E. Ushakov, eds. **Reliability Handbook.** New York: Holt, 1970. 391 pp.

> This volume presents a comprehensive review of theory, followed by problems on optimal use.

Radar

Barton, David K., and H. R. Ward, eds. **Handbook of Radar Measurement.** Englewood Cliffs, N.J.: Prentice-Hall, 1969. 426 pp.

> Eight chapters deal with such aspects of the subject as angular and range measurement in noise and target-induced error. The appendixes contain basic tables.

Skolnik, Merrill I., ed. **Radar Handbook.** New York: McGraw-Hill, 1970. Variously paged.

> The handbook covers all aspects of the subject in its 38 chapters. It includes both theory and application, with more emphasis on application.

Walton, J. D., Jr., ed. **Radome Engineering Handbook: Design and Principles.** New York: Marcel Dekker, Inc., 1970. 592 pp.

> This work thoroughly discusses the electrical, chemical, and mechanical features of radomes as well as includes the special problem of rain erosion.

Safety

Factory Mutual System. **Handbook of Industrial Loss Prevention.** 2d ed. New York: McGraw-Hill, 1967. 912 pp.

The book gives details of methods to protect industrial plants and processes from damage by fires, lightning, and earthquakes. It discusses sprinkler systems, handling of flammable materials, and the like.

Hammer, Willie, ed. **Handbook of System and Product Safety.** Englewood Cliffs, N.J.: Prentice-Hall, 1972. 351 pp.

Besides the scope indicated by its title, this book includes applications of safety to specific areas, such as mining, motor vehicles, railroads, and general industry. It is oriented more toward systems safety than product safety. Other topics include legal liability, basic concepts of hazards, and operating analyses.

Handbook of Laboratory Safety. 2d ed. Edited by Norman V. Steere. Cleveland: CRC Press, 1971. 854 pp.

This handbook contains a wide range of topics, including protective equipment, ventilation, fire hazards, chemical reactions, and toxic, radiational, electrical, and biological hazards.

National Fire Protection Association. **Fire Protection Handbook.** 14th ed. Boston: The Association, 1976. 1,296 pp.

Beginning with the basic types of fire losses, the book proceeds to a discussion of fire hazards of various materials, fire protection in buildings, fire-fighting equipment, extinguishing agents, and so forth.

Sax, N. Irving, ed. **Dangerous Properties of Industrial Materials.** 4th ed. New York: Van Nostrand Reinhold, 1975. 1,216 pp.

The main portion of the book is the section listing over 12,000 industrial and laboratory materials. It gives the formula, physical description, types of hazards involved, and countermeasures to use for each. Other sections discuss topics such as pollution control, industrial fires, radiation damage, and shipping regulations.

Stress Analysis

Griffel, William, ed. **Handbook of Formulas for Stress and Strain.** New York: Ungar, 1966. 418 pp.

This compilation of formulas, diagrams, charts, and tables gives the important data on the subject. It discusses types of objects such as plates and columns as well as kinds of phenomena such as torsion and vibration.

Hetenyi, Miklos I., ed. **Handbook of Experimental Stress Analysis.** New York: Wiley, 1950. 1,077 pp.

> Following a description of different types of gauges and measuring techniques are chapters on various kinds of stresses, theoretical aspects of elasticity, and so forth.

Technical Writing

Jordan, Stello, ed. **Handbook of Technical Writing Practices.** New York: Wiley-Interscience, 1971. 2 vols.

> This volume is aimed at improving the quality of technical writing of a wide range of users, both commercial and military. Its 32 chapters deal with diverse topics, as exemplified by equipment instruction manuals, technical reports, sales literature, and technical films.

Timber

Timber Engineering Company. **Timber Design and Construction Handbook.** New York: McGraw-Hill, 1956. 622 pp.

> Among the topics discussed are roof trusses, plywood, fabrication, and exterior structures.

American Institute of Timber Construction. **Timber Construction Manual.** 2d ed. New York: Wiley-Interscience, 1974. 799 pp.

> New material added since the first edition includes design stresses, new lumber sizes, revised design procedures for wood structural elements and fastenings, plus standard data on a wide range of topics.

Transportation Engineering

Society of Automotive Engineers. **SAE Handbook.** New York: The Society. 1913-   . [Annual]

> This compilation of SAE Standards and Recommended Practices for all aspects of vehicle design and manufacture includes materials, components, and the like.

Urban Planning

Claire, William H., ed. **Handbook on Urban Planning.** New York: Van Nostrand Reinhold, 1973. 416 pp.

> This work explains all aspects of urban planning, including the engineering, architectural, and commercial. It considers housing, business, and transportation facilities.

Vacuum Engineering

Steinherz, H. A., ed. **Handbook of High Vacuum Engineering.** New York: Van Nostrand Reinhold, 1963. 358 pp.

> The handbook has chapters on general topics (behavior of gases at low pressure, materials of construction, and so forth) as well as parts of vacuum pumps (seals, baffles, valves, and the like).

Vibrations

Harris, Cyril M., and C. E. Crede, eds. **Shock and Vibration Handbook.** New York: McGraw-Hill, 1961. 3 vols.

> In this comprehensive work, volume 1 covers basic theory and measurements, volume 2 deals with data-analysis testing and methods of control, and volume 3 is on engineering design and environmental conditions.

Water Supply

American Water Works Association. **Water Quality and Treatment: A Handbook of Public Water Supplies.** 3d ed. New York: McGraw-Hill, 1971. 654 pp.

> This book discusses basic aspects of water treatment, such as aeration, taste, and odor control, and filtration as well as related topics such as management of plant residues and handling of chemicals.

White, George C., ed. **Handbook of Chlorination: For Potable Water, Wastewater, Cooling Water, Industrial Processes, and Swimming Pools.** New York: Van Nostrand Reinhold, 1972. 744 pp.

> This volume has chapters on various applications of chlorination, as well as on equipment for the process. It gives the subject thorough coverage.

See also **Hydrology**

X-Rays

Kaelble, E. F., ed. **Handbook of X-Rays: For Diffraction, Emission, Absorption, and Microscopy.** New York: McGraw-Hill, 1967. Variously paged.

> The chapters are grouped under the headings of Fundamentals, Diffraction, Determination of Crystal Structure, Emission Spectroscopy, Absorption Methods, and Microradiography.

CHAPTER 7

Guides to the Literature

General guides to the literature vary tremendously in scope and style. Some cite only a few distinguished sources, while others list many sources indiscriminately. Usually the greater the number of entries, the less chance there is for carefully prepared annotations for each item; yet annotations greatly increase the value of the book. A mere citation, without some hint of the item's worth to the reader, is of questionable value. This is especially true because guides are ordinarily intended for the person not strongly grounded in the writings of the subject.

A guide should also have descriptive analyses of the general use and worth of each **class** of work as a whole.

SELECTED EXAMPLES

General

Winchell, Constance M., comp. **Guide to Reference Books.** 8th ed. Chicago: American Library Association, 1967. 741 pp. and supplements.

> This well-known work lists reference books under broad categories, such as humanities, science, and so forth, with more detailed subject breakdowns under the main divisions. The items are annotated and coverage is thorough. There are author, subject, and title indexes. Three supplements, edited by Eugene P. Sheehy, have updated the basic volumes for the period 1965-1970.

General—Science and Engineering

Grogan, Dennis. **Science and Technology: An Introduction to the Literature.** 2d rev. ed. Hamden, Conn.: Linnet, 1973. 254 pp.

> This full treatment of the literature is arranged by type (patents, handbooks, and so on) and includes comments on the characteristics of each. It evaluates thousands of items in a running commentary rather than in formal citations and lists.

Houghton, Bernard. **Technical Information Sources: A Guide to the Patents, Standards, and Technical Reports Literature.** 2d rev. ed. Hamden, Conn.: Shoe String: 1973. 119 pp.

> This volume has five chapters on patents, including their nature and bibliographic control. There are three chapters on standards, and two more on technical reports. Considerable attention is devoted to these specific types of technical literature.

Lasworth, Earl James. **Reference Sources in Science and Technology.** Metuchen, N.J.: Scarecrow Press, 1972. 305 pp.

> Lasworth describes the features of different types of technical and scientific literature (such as handbooks, encyclopedias, translations, periodicals), then lists selected samples, without annotations. One chapter includes guidance on the use of a card catalog. The last chapter discusses the proper style of bibliographic references and the use of style manuals.

Malinowsky, H. Robert. **Science and Engineering Reference Sources: A Guide for Students and Librarians.** 2d ed. Littleton, Colo.: Libraries Unlimited, 1976.

> This book devotes a chapter each to the literature of many disciplines (physics, chemistry, engineering, medicine, geology, biology, mathematics, and general science). It cites hundreds of examples, many of which are annotated, and has an author-title index.

Parsons, Stanley Alfred James. **How to Find Out about Engineering.** New York: Pergamon, 1972. 271 pp.

> Introductory chapters cover topics such as engineering careers and the basics of libraries, followed by chapters on different types of materials (handbooks, periodicals, etc.) Following chapters cover specific branches of engineering, such as mechanical, civil, and mining. Selected examples of literature are described. There are name and subject (plus title) indexes.

Walford, A. J. **Guide to Reference Material.** Vol. 1 **Science and Technology.** 3d ed. New York: Bowker, 1973. 624 pp.

> This volume lists several thousand books and reference sources under broad categories, such as geology, engineering, and chemical industry, as well as general technical works. All items are annotated, and there is an author-title index.

Agricultural Engineering

American Society for Engineering Education. Engineering School Libraries Division. **Guide to Literature on Agricultural Engineering.** Compiled by Elizabeth P. Roberts. Washington, D.C.: The Society, 1971. 19 pp.

> This pamphlet consists of a listing, without annotations, of selected recommended works in eleven categories, including guides to the literature, dictionaries, serials, and major research centers.

Astronautics

Fry, Bernard M., and F. E. Mohrhardt, eds. **A Guide to Information Sources in Space Science and Technology.** New York: Wiley, 1963. 579 pp.

> Besides listing traditional reference books, this guide cites thousands of technical reports and periodical articles on specific topics, such as lunar trajectories, metals, and propellants. All items are annotated. It has author, title, and subject indexes.

Atomic Energy

Anthony, L. J. **Sources of Information on Atomic Energy.** Elmsford, N.Y.: Pergamon, 1966. 245 pp.

> This thorough treatment of the literature lists national sources of information, arranged by country, and then gives international sources. Following these sources are several chapters on published literature of various types that pertain to five aspects of the subject—such as high energy physics, and ionizing radiation.

Biology

Bottle, R. T., and H. V. Wyatt, eds. **The Use of Biological Literature.** Hamden, Conn.: Archon/Shoe String, 1966. 286 pp.

> Various types of literature, as well as the literature of various disciplines in the life sciences (zoology, microbiology, etc.), are here discussed in a collection of articles. Thousands of titles are

mentioned, and most of them are either annotated or described. There is a subject index.

Chemical Engineering and Chemistry

Bottle, R. T., ed. **The Use of Chemical Literature.** 2d ed. Hamden, Conn.: Archon/Shoe String, 1969. 294 pp.

> Chapters cover either types of literature (e.g., abstracting services or dictionaries) or major topics important in chemistry (e.g., nuclear chemistry or inorganic chemistry). Two chapters discuss how to use libraries, and search techniques.

Mellon, M G. **Chemical Publications: Their Nature and Use.** New York: McGraw-Hill, 1965. 324 pp.

> A familiar text to many users, this work not only gives the characteristics, with examples, of many types of literature, but has a lengthy chapter on the techniques of searching the chemical literature.

Smith, Julian F., and T. E. Singer, eds. **Literature of Chemical Technology.** Advances in Chemistry Series no. 78. Columbus, Ohio: American Chemical Society, 1968. 732 pp.

> Forty very thorough chapters list thousands of citations on various aspects of the literature, such as refractories, the cosmetics industry, carbon black, and resins and plastics. The book is based on two ACS Symposia held in 1963.

Woodburn, Henry M. **Using the Chemical Literature.** New York: Marcel Dekker, Inc., 1974. 302 pp.

> Aimed at practicing chemists as well as at college-level students, this book is arranged by type of literature as well as partially by subject (e.g., organic chemistry and spectroscopy). It includes a chapter on retrospective searching.

Computers

Carter, Ciel M. **Guide to Reference Sources in the Computer Sciences.** New York: Macmillan Information, 1974. 237 pp.

> This volume includes organizations as well as primary literature and reference materials. Around 900 Sources are described. A very thorough work.

Electrical Engineering

Burkett, Jack, and P. Plumb. **How to Find Out in Electrical Engineering.** Elmsford, N.Y.: Pergamon, 1967. 234 pp.

> The authors use the Universal Decimal Classification for the arrangement of sources, and devote chapters to such subjects as electronics, power generation, data processing, etc. They also devote chapters to forms of literature, such as encyclopedias, tables, and periodicals.

Mathematics

Parke, Nathan Grier, III. **Guide to the Literature of Mathematics and Physics, Including Related Works on Engineering Science.** 2d rev. ed. New York: Dover, 1958. 436 pp.

> Part I of this comprehensive work discusses general topics, such as study habits, literature searching, and types of reference books. Part II lists thousands of texts for specific subjects, such as conformal mapping and piezoelectricity, as well as types of literature (periodicals, dictionaries, and so on). The work is in need of updating.

Pemberton, John E. **How to Find Out in Mathematics: A Guide to Sources of Information.** 2d ed. Elmsford, N.Y.: Pergamon, 1970. 220 pp.

> This guide reviews the characteristics of various types of material (dictionaries, periodicals, theses, and so forth) and cites examples of each. Later chapters discuss the literature of specific fields, such as probability, operations research, and actuarial science. The book has an author-title index.

Mechanical Engineering

Houghton, Bernard. **Mechanical Engineering: The Sources of Information.** Hamden, Conn.: Shoe String, 1970. 311 pp.

> This work not only discusses the different kinds of literature but also describes other sources of information, such as government agencies, trade associations, and research institutions. It lists hundreds of examples, most of which are not annotated. This comprehensive guide has an author-title index.

Siddall, James N. **Mechanical Design: Reference Sources.** Toronto: University of Toronto Press, 1967. 156 pp.

> This book cites hundreds of recommended publications, arranged

under such headings as fluid-flow machinery, adhesives, and value engineering. No annotations are given.

Metallurgical Engineering and Metals

Gibson, Eleanor B., and E. W. Tapia, eds. **Guide to Metallurgical Information.** SLA Bibliography, no. 3. 2d ed. New York: Special Libraries Association, 1965. 222 pp.

> This work lists pertinent information centers as well as types of literature of interest to metallurgists. All cited items are annotated. It has several indexes, including ones for personal authors, subjects, titles, and organization names.

Hyslop, Marjorie R. **A Brief Guide to Sources of Metals Information.** Washington, D.C.: Information Resources Press, 1974. 180 pp.

> This guide describes many types of literature of interest and also gives a directory of information sources and guidelines on the use of libraries.

Mining Engineering and Minerals

Kaplan, Stuart R., ed. **Guide to Information Sources in Mining, Minerals, and Geosciences.** New York: Wiley, 1965. 599 pp.

> Part I cites important sources organizations on a geographical basis, indicating fields of study, publications, and so forth. Part II gives major types of literature for each of 20 disciplines, such as geology, mining, and physics.

Physics

Whitford, Robert N. **Physics Literature: A Reference Manual.** 2d ed. Metuchen, N.J.: Scarecrow, 1968. 272 pp.

> This book, arranged by kind of physics literature, such as historical, biographical, and mathematical, lists thousands of examples, with concise comments for many of them. It has an author and subject index.

Yates, Bryan. **How to Find Out About Physics: A Guide to Sources of Information.** Elmsford, N.Y.: Pergamon, 1965. 175 pp.

> Yates arranges sources of information by the decimal system, collected in chapters under such broad headings as optics, heat, and atomic physics. Some chapters discuss types of literature (periodicals, documents, etc.). The subject-title index is selective.

CHAPTER 8

Histories

A history of a type of engineering, such as mining, cannot be rated as an absolute must for most engineers, yet somewhere along the line they would do well to read the history of their discipline. A history can give a good perspective of the present, and this perspective can show up current momentary perturbations for what they are. It can also show up past mistakes which can then, one hopes, be more easily avoided now.

Not all histories are so obliging as to set forth these mistakes clearly, and some are so bland that the reader will look in vain for any clear-cut observations and conclusions. Others are mostly pictorial.

The older the field, of course, the longer the work is apt to be. A history of computer science would probably not yet fill a large book, although older workers in that field have already written a number of useful periodical articles. On the other hand, books written in the sixteenth century on metallurgy and mining are now curiosity pieces, worthless as current aids but priceless because of what they tell us of the history of that discipline as well as the state of the art of that era. The earliest volume of one of the world's first scientific journals, the publication of the Academy of Sciences in Paris, was issued in 1699, and by following its columns through the centuries, one can get a tremendous sense of the interests and developments of the times. Another rich source concerning the developments of a science is the personal correspondence of great scientists and engineers.

The history of engineering and science has become a discipline of its own. Some valuable library collections are available on the subject, college courses are offered in it, and books are being written on small segments of the vast body of data available.

SELECTED EXAMPLES

General—Engineering

Ferguson, Eugene S., ed. **Bibliography of the History of Technology.** Cambridge, Mass.: M.I.T., 1968. 347 pp.

> This impressive, annotated compilation of works arranged in broad categories such as manuscripts and biographies, as well as by a dozen or so subjects (civil engineering, materials and processes, and so on), has an author index and emphasizes literature that contains bibliographies.

Rapport, Samuel, and Helen Wright, eds. **Engineering.** New York: New York University Press, 1963. 378 pp.

> This collection of essays covers the period from ancient times to the present. It ranges from a discussion of da Vinci's works to the uses of atomic energy. This is an interesting compilation.

Specific Disciplines

Dennis, William H. **A Hundred Years of Metallurgy.** Chicago: Aldine, 1963. 342 pp.

> The nine chapters of this volume are divided according to types of metals or by processes. Each chapter covers a wide time period. The work includes a great deal of detailed information and has a name and a subject index.

Shiers, George. **Bibliography of the History of Electronics.** Metuchen, N.J.: Scarecrow, 1972. 323 pp.

> More than 1,800 items, arranged in a classification system, cover the historical highlights of electronics, including telecommunication.

Straub, Hans. **History of Civil Engineering: An Outline from Ancient to Modern Times.** Cambridge, Mass.: M.I.T., 1952. 258 pp.

> This history covers thousands of years and discusses techniques and problems encountered in each era. It is well illustrated and has separate indexes for subjects, personal names, and place names.

CHAPTER 9

Biographical Information

Biographical reference works about scientists and engineers vary widely. Some include age, education, family description, occupational history, and sometimes important writings. Others state only the present address, education, and perhaps the subject's specialty of job title.

One problem with such works is that in some disciplines there are no frequently issued compilations, and in others the only up-to-date biographical sources are extremely sketchy. The membership lists of scientific or technical societies are completely democratic, for example, and include all members, thus citing many lesser-known people who are difficult to learn about through other sources. But the amount of information given is minimal.

One biographical reference often overlooked is the author index for periodical articles. If the person has written recently, the journal carrying the article usually includes at least the current address and job title of the author, and often much more.

Not finding a person listed in a biographical source doesn't indicate unimportance in the field, since methods of selection are so erratic that many worthy names are overlooked. Conversely, lengthy sketches are no guarantee of professional ability, since some compilers allow entrants to make their entries about as long as they desire.

SELECTED EXAMPLES

General—Science and Engineering

American Men and Women of Science. Physical and Biological Sciences Section. 12th ed. Ann Arbor, Mich.: Bowker, 1971-1973. 6 vols.

> This set contains over 145,000 biographies in 600 areas of science. Besides the usual information, it includes the area of specialization, society memberships and so on. There is a separate necrology. Another set of volumes covers the social and behavioral sciences.

Dictionary of Scientific Biography. Edited by C. G. Gillispie. New York: Scribner, 1970-

> This impressive, carefully prepared compilation is expected to fill 12 volumes (and have 5,000 biographies) plus an index volume when it is completed. It is cosponsored by the National Science Foundation and the American Council of Learned Societies, excludes living scientists, and extends back to antiquity. Twelve volumes had been published by 1976.

Engineers Joint Council. **Engineers of Distinction.** 2d ed. New York: The Council, 1973. 400 pp.

> Limited to the winners of national awards, and to officers, directors, and chief staff officers of national engineering societies, this book lists around 5,000 individuals. It gives standard biographical information and is quite selective in choice of nominees for inclusion.

Ho, James K. K., comp. **Black Engineers in the United States.** Washington, D.C.: Howard University Press, 1974. 281 pp.

> This work alphabetically lists 1,500 names, an estimated 20 percent of the total number of such engineers. It has indexes by state and by major disciplines. It includes some students.

McGraw-Hill Modern Men of Science. New York, McGraw-Hill, 1966-1968. 2 vols.

> This work is limited to fewer than 1,000 invited scientists. Each article includes a hand-drawn sketch of the scientist as well as an account—often several pages long—of the person's career. It has a name and subject index.

Who is Publishing in Science: An International Directory of Research and Development Scientists. Philadelphia: Institute for Scientific Information, 1967-   [Annual]

This volume lists authors alphabetically, showing their current work affiliation. The names are from those whose publications have been indexed in **Science Citation Index.** Over 250,000 scientists and technical authors are listed.

Who's Who in Engineering. 9th ed. New York: Lewis Historical Publishing Co., 1964. 2,198 pp.

Although out of print, this book is still a useful source for data on thousands of engineers. Restricted to those with 15 years of experience, it gives standard biographical information.

Who's Who in Science in Europe. 2d ed. Guernsey, British Isles: Hodgson, 1971. 4 vols.

More than 40,000 East and West European scientists are included. This work is not just a revision of the first edition but is a new compilation.

World Who's Who in Science: A Biographical Dictionary of Notable Scientists from Antiquity to the Present. Chicago: Marquis, 1968. 1,855 pp.

Besides the usual personal information, this dictionary cites the major contributions of around 30,000 scientists.

Computers and Data Processing

Who's Who in Computers and Data Processing. 6th ed. Newtonville, Mass.: Berkeley Enterprises; New York: New York Times Book Co./Quadrangle Books [in prep.]

This work gives standard biographical information about systems analysts, programmers, data processing managers, and other computer personnel.

CHAPTER 10

Directories and Yearbooks

A directory is a listing of individuals, companies, products, or places organized according to specific rules or restrictions. A telephone book is obviously limited to certain geographical boundaries, whereas other lists of names may be organized according to professional title, age, affiliation, and so on.

In the case of organizations, an example of one of the most useful directories, **The Encyclopedia of Associations,** gives political, social, professional, religious, and economic associations, as well as information about the group's main office, its purpose, size, publications, and so on. Some directories are international in scope and still manage to issue new editions quite frequently (see **World of Learning**).

Other directories of organizations may be limited to a group of great importance to many engineers—manufacturers, for example. Certain directories list thousands of manufacturers all over the U.S. with no restriction as to products, while others limit themselves to those in a certain geographical area or a particular activity, such as mining or electronics. For more information about directories of manufacturers and their products, see Chapter 18. Biographical sources, as described in Chapter 9, could also be listed as a type of directory.

The yearbook is another type of publication that in some ways is similar to the directory and in other ways to an annual review. Yearbooks are general summaries, including statistics, of the highlights of a particular subject field, industry, or area of research;

so in a sense they are also related to items discussed in Chapter 19—statistical sources.

At the close of this chapter, a selected listing of yearbooks is given to illustrate the breadth of information available through yearbooks in general. Many other equally worthwhile publications could have been selected.

SELECTED EXAMPLES

General—Directories

Encyclopedia of Associations. 9th ed. Detroit: Gale Research Company, 1975. 3 vols.

> In volume 1 over 14,000 national associations in the United States are arranged by categories, such as educational, religious, technical, and fraternal. There is also a key-word index. Volume 2 lists the groups geographically and contains an index of executives. It also gives subject interests, number of members, publications, and so forth. Volume 3 is in a looseleaf format for adding quarterly supplements of new associations.

Research Centers Directory. 5th ed. Edited by Archie M. Palmer. Detroit: Gale Research Company, 1975. 1,056 pp.

> The book lists some 6,000 centers, arranged by broad subject categories, and gives the name of the director and the principal fields of research. It has alphabetical indexes of center name, subjects of research, and personal names. Updated by a publication entitled **New Research Centers**.

World Guide to Scientific Associations. New York: Bowker, 1974. 481 pp.

> This guide is a directory to over 10,000 associations (from 134 countries) that are involved in science and research. The associations are listed by continent, then by country, then alphabetically by title. There is also a detailed subject index.

World of Learning. London: Europa Publications, 1947-   [Annual]

> This truly international directory gives descriptions of colleges and universities all over the world, as well as names of faculty members and administrators, and then gives the same sort of information for museums, libraries, cultural and educational institutions, and the like. It is a remarkably comprehensive work.

Specific—Directories

American Society for Testing and Materials. **Directory of Testing Laboratories, Commercial-Institutional.** Philadelphia: The Society, 1969. 28 pp. (ASTM STP 333A)

> The directory lists laboratories geographically and gives the location and the type of tests done (a special code represents the types). There is also an alphabetical index of laboratories.

Cass, James, and Max Birnbaum, eds. **Comparative Guide to Science and Engineering Programs.** New York: Harper & Row, 1971. 1,165 pp.

> The guide describes the programs in natural sciences, mathematics, and engineering that are offered by 2,500 departments in selected colleges and universities. It lists faculty size, types of courses, grade requirements, outlook for graduate study, and so on. Several tables with information such as number of degrees offered or number of students entering graduate study are arranged by school.

Directory of Engineering Societies and Related Organizations. New York: Engineers Joint Council, 1974. 178 pp.

> This work is a compilation of data on more than three hundred national, regional, and international organizations related to engineering.

Directory of Engineers in Private Practice, 1970-71. Washington, D.C.: National Society of Professional Engineers, 1970. 335 pp.

> Published by the Professional Engineers in Private Practice section of NSPE, this work lists members alphabetically, then by states, then by their firms. There is a subject index to firms that have certain engineering specialties.

Directory of Scientific Directories. Compiled by Anthony P. Harvey. Guernsey, British Isles: Hodgson, 1969. 272 pp.

> This international guide to scientific directories includes the topics of engineering, medicine, manufacturing, and so on. It is arranged geographically, with items annotated, and cites over 1,600 publications.

E/MJ International Directory of Mining and Mineral Processing Operations. New York: McGraw-Hill, 1968-   [Annual]

> This directory of active mineral producers—on an international

basis—includes names of companies, contractors, and executives, as well as companies arranged by type of minerals produced.

Energy Directory. New York: Environment Information Center, 1974. 418 pp.

This directory lists data from over 3,000 organizations—including the names of 12,000 officials—that are involved in energy projects and activities. It includes government and private groups, and has indexes by organization, subject, and geography.

Guide to American Scientific and Technical Directories. Edited by Bernard Klein. 2d ed. Rye, N.Y.: 1975. 350 pp.

Over 2,500 directories covering the physical sciences, industrial and technical areas, and the social sciences are listed here. Name and source are given for each directory. There is a subject index.

Industrial Research Laboratories of the United States. 14 ed. New York: Bowker, 1975. 585 pp.

This book cites over 6,600 laboratories within 3,115 companies and gives the name of the company, key personnel, the field and educational level of the research laboratory staff, and the research interests. There are subject, personnel, and geographic indexes.

Scientific, Technical and Engineering Societies Publications in Print, 1974-1975. Edited by James M. Kyed and James M. Matarazzo. New York: Bowker, 1974. 223 pp.

This volume lists the publications (print and nonprint materials) of over 150 scientific and engineering societies. There are indexes by keyword and authors or editors.

Scientific, Technical, and Related Societies of the United States. 9th ed. Washington, D.C.: National Academy of Sciences, 1971. 213 pp.

This work describes the size, headquarters' location, names of publications, areas of interest, and so forth of over 500 societies. It has a detailed subject index.

World Directory of Environmental Research Centers. 2d ed. Edited by William K. Wilson and others. New York: Bowker, 1974. 448 pp.

This volume provides data on nearly 5,300 organizations involved in environmental research in over 100 countries. Staff size and activities of the group, both government and private, are described.

DIRECTORIES AND YEARBOOKS 95

Yearbooks

Aerospace Industries Association of America. **Aerospace Year Book.** New York: Spartan, 1945-   [Annual]

> This publication reviews important technical and commercial developments and is profusely illustrated.

Jane's All the World's Aircraft. New York: F. Watts, 1909- [Biennial]

> This well-illustrated yearbook of the aviation industry describes features of the new aircraft of the world and is arranged by countries. It also has separate sections describing rockets and space vehicles, military missiles, engines, drones, and sailplanes.

McGraw-Hill Yearbook of Science and Technology. New York: McGraw-Hill, 1962-   [Annual]

> This yearbook provides updating information for the **McGraw-Hill Encyclopedia of Science and Technology,** with the largest portion devoted to an alphabetically arranged series of articles on the year's developments. One segment is devoted to selected scientific photographs of the year, and another section contains articles on selected topics of more than average interest.

U.S. Bureau of Mines. **Minerals Yearbook.** Washington, D.C.: Superintendent of Documents, 1933-   [Annual]

> This review of the world's minerals industry is usually in four volumes that are devoted to metals, minerals, and fuels; area reports (domestic); and area reports (international). They discuss prices, production, consumption, and the like.

CHAPTER 11

Annual Review Series

Sometime during the 1940s, a type of literature that aimed to present an annual review of the important developments in a particular scientific discipline became popular. Each article in the review was written by an expert who would go through and review the literature of that year, analyze other developments in the field, and then summarize all major trends and important results. Other series aimed solely at presenting outstanding developments, often in long, scholarly papers.

Although these two types of series (usually titled "Progress in..." or "Advances in...") are almost alike in content—there is little of the annual review feature left, and the papers tend to be lengthy, often with long bibliographies and many citations. They have proliferated enormously, and have also become enormously important.

A large problem is that the **contents** of many of these series have not been indexed anywhere, and as a consequence the individual papers are often not well known. Yet many of them contain major reviews of the literature on an important, even if narrow, topic. Usually engineers find one or two series that interest them and keep track of these volumes, although they are often issued very irregularly, and a lapse of two or three years between volumes is not uncommon.

Fortunately, a new service (**Index to Scientific Reviews**) designed to index reviews, whether in periodical articles or annual review series, began recently and should go far to alleviate the problems of indexing annual reviews.

In the meantime, except for those few remaining annual reviews that actually still do examine the year's developments, engineers must turn elsewhere for a state-of-the-art survey, such as an article or report that has as its goal a review of what is now feasible in a particular field (see Part III).

SELECTED EXAMPLES

Because of the great number of annual review series in existence, only a few of those of greatest interest are listed here. Some have been in existence for over a decade, others have issued only two or three volumes to date. Only title and publisher are given.

Advances in . . . Series

Advances in Analytical Chemistry and Instrumentation (Wiley)
Applied Mechanics (Academic)
Biochemical Engineering (Springer-Verlag)
Biological and Medical Engineering (Academic)
Biomedical Engineering and Medical Physics (Wiley)
Chemical Engineering (Academic)
Communication Systems (Academic)
Computers (Academic)
Control Systems (Academic)
Cryogenic Engineering (Plenum)
Ecological Research (Academic)
Electrochemistry and Electrochemical Engineering (Wiley)
Electronics and Electron Physics (Academic)
Environmental Sciences (Wiley)
Geophysics (Academic)
Heat Transfer (Academic)
Hydroscience (Academic)
Machine Tool Design and Research (Pergamon)
Materials Research (Wiley)
Microwaves (Academic)
Nuclear Science and Technology (Academic)
Petroleum Chemistry and Refining (Wiley)
Plasma Physics (Wiley)
Polymer Science (Springer-Verlag)
Space Science and Technology (Academic)
Vacuum Science and Technology (Pergamon)
Water Pollution Research (Pergamon)
X-Ray Analysis (Plenum)

Annual Review of . . . series

Annual Review of Automatic Programming (Pergamon)
  Fluid Mechanics (Annual Reviews, Inc.)
  Materials Science (Annual Reviews, Inc.)
  Nuclear Science (Annual Reviews, Inc.)

Progress in . . . series

Progress in Aeronautical Sciences (Pergamon)
  Applied Materials Research (Gordon)
  Ceramic Science (Pergamon)
  Cryogenics (Academic)
  Dielectrics (Academic)
  Heat and Mass Transfer (Pergamon)
  High Temperature Physics and Chemistry (Pergamon)
  Material Science (Pergamon)
  Oceanography (Pergamon)
  Polymer Science (Pergamon)
  Quantum Electronics (Pergamon)
  Semiconductors (Wiley)
  Solid Mechanics (American Elsevier)

Index to Scientific Reviews. Philadelphia: Institute for Scientific Information, 1975-   [Semiannual; annual cumulations]

This work provides an index for review articles found in more than 2,700 periodicals and annual review publications. Its indexes are by author, title words, and organizations.

Part III
Periodicals and Technical Reports

This section discusses only two types of literature—periodicals and technical reports—which are two of the most important sources of up-to-date, detailed information that an engineer will have occasion to use. In sheer volume, as well as in value to the engineer, they stand distinct from many of the other types of information sources.

CHAPTER 12

Periodicals

One of the most important sources of information for the engineer is the technical journal or periodical. Periodicals are indispensable for many reasons, of which one of the most important is that they present data that is relatively new. Another characteristic of periodicals is that their articles are usually quite specific, often must more so than a book on the same subject.

The importance of technical periodicals may be seen from the **World List of Scientific Periodicals** which lists over 60,000 different periodicals published from 1900 to 1960. (see also Chapter 1 of this book). Naturally, not all these publications are still in existence, but the very number of them indicates the significance of this type of literature. The periodical as a form of publication has a long history going back more than 200 years. The range of periodicals is quite broad, with those of a topical, news-item nature at one end of the spectrum and those of a scholarly, more technical nature at the other. Periodicals may be all in English, all in a foreign tongue, or a mixture. Some are simply translations of others. Sponsors range from commercial sources to trade associations or engineering societies.

As mentioned in Chapter 1, the scope of many journals is currently being broadened because of increased demands for projects and products that are responsive to the social and economic needs of our times. Fewer journals than ever deal 100 percent with the strictly technical aspects of a branch of engineering, and many discuss ecology, federal safety standards, the ethics of research in military weapons, and other topics that would have been much harder to find in these same journals even 15 years ago.

Today's more rapid technical developments are also due in part to the success of many periodicals in getting out important news faster, although others pursue their apparently unhurried life with little regard for the "newsworthiness" of an item. However, it is noteworthy that often a newspaper's source of information for a key feature is a technical journal. Some prestigious journals have first call on the announcement of any new data because they will not knowingly print information that has already been published. They take pride in being known—sometimes internationally—as sources of first disclosures of new technical developments.

Owing to the great number of periodicals and the subjects they cover, reference tools are necessary to help one find any wanted information. To find a particular article within a given journal, for example, requires the use of periodical indexing and abstracting services, which are described in Chapter 13.

There are also directories that list journals currently being published, giving prices, publishers, name of the indexing service, frequence of publication, and so on. Usually there is also a subject index.

Quite often, however, the problem is one of trying to locate an issue of a periodical not carried by the library one normally uses. The tool to apply then is called a **union list**, meaning a list showing the periodicals held by a group (or union) of libraries that have consolidated their collection information into one list. The most detailed union lists cover the holdings of major college and public libraries in the United States, while smaller lists are restricted to one region, state, city, or even to the various branches of one library system. Others are restricted by subject. One tool (**New Serial Titles**) is confined to periodicals that are more or less current, thus supplementing the major union list (see **Union List of Serials . . .**), which apparently will never be updated, so expensive has it become.

SELECTED EXAMPLES

Bibliographic Guide for Editors & Authors. Columbus, Ohio: Chemical Abstracts Service, 1974. 362 pp.

This bibliography lists nearly 28,000 titles of technical and scientific journals, of which around two-thirds are still being published. It indicates which are indexed by the three groups sponsoring the publication (Engineering Index, Inc., BioSciences Information Service of Biological Abstracts, and CAS), and gives title abbreviations, International Standard Serial Numbers (ISSN) and/or CODEN.

British Union-Catalogue of Periodicals, Incorporating World List of Scientific Periodicals: New Periodical Titles. London: Butterworth, 1964-   [Quarterly, annual cumulations]

This work cites new periodicals published since 1960, thus updating the fourth (and last) edition of **World List of Scientific Periodicals.** It also gives the holdings of selected British libraries.

New Serial Titles: A Union List of Serials Commencing Publication after Dec. 31, 1949. Washington, D.C.: Library of Congress, 1950- [Monthly, quarterly and annual cumulations]

This work updates the **Union List of Serials** but is restricted to the United States. It includes holdings of chosen libraries. Five- and ten-year cumulations are also available

Ulrich's International Periodicals Directory. New York, Bowker, 1932-   [Biennial]

The 1975/76 edition lists over 57,000 serials under about 200 subject categories. It gives full title, sponsoring organization or publisher, editors, cost, indexing service, and so forth of each journal and whether book reviews are included. It is a well-known tool, found in almost all libraries.

Union List of Serials in Libraries of the U.S. and Canada. 3d ed. New York, Wilson, 1965. 5 vols.

Citing around a third of a million periodicals, this list gives bibliographic data about the place and name of the publisher, date commenced, and title changes of holdings of hundreds of university and public libraries in the U.S. and Canada. Updated by **New Serial Titles.**

World List of Scientific Periodicals, Published in the Years 1900-1960. 4th ed. Edited by Peter Brown and G. B. Stratton. London: Butterworth, 1963-1965. 3 vols.

Listing title, commencement date, and place of publication of holdings of selected British libraries, this work covers nearly 60,000 titles. Updated now by **British Union-Catalogue of Periodicals.**

CHAPTER 13

Periodical Indexing and Abstracting Services

Thousands of technical articles are published each year in the professional journals. The output is so prodigious that there is no possible way that one could scan—let alone read—the thousands of journals that might occasionally include an article of importance on a particular topic of interest to an individual. This fact has led to the creation of an impressive number of indexing and abstracting services that provide current information on what is being published.

An **indexing** service usually cites the article title, author, subject, and bibliographic data. An **abstracting** service provides in addition summaries of the information in the articles cited.

Abstracts range from brief statements (indicative abstracts) of the article's highlights to detailed accounts (informative abstracts) that cite minute facts, such as the exact materials used and their proportions.

Both types of services are often known as secondary services because they reprocess original material, or primary sources. In general, indexing services cover a broader subject field than those that provide abstracts. Both types are indispensable to the engineer, since there is no other way that the individual can hope to be aware of most of the important new developments in his field.

The rate at which the technical literature is expanding has led to what has become well-known as the information explosion. For example, **Chemical Abstracts,** which deals with pure and applied chemistry and related fields, published its five-millionth abstract in December, 1971, just a bit more than three years after handling its four-millionth abstract. In contrast, it took from 1907 to 1938 for **Chemical Abstracts** to reach its first one-millionth abstract.[1]

Indexing services indicate the scope of their coverage of periodicals and usually list those titles they deal with regularly. Treatment can vary tremendously, ranging from some services that index everything in each journal (including news items, obituaries, cost, and statistical data) to those that cite only articles of outstanding significance.

The manner in which they are arranged varies considerably. Some list articles under alphabetically arranged subject headings, more or less as a card catalog. Others organize them according to a special classification system of their own and include a subject index (in alphabetical order) to this scheme. Still others, fortunately not many, use the dubious method of arranging articles by date of issue, with subject indexes to help locate items. A good service is also likely to have special indexes, such as one by patent numbers in an index for patents, or one based on structural patterns in an index for chemistry.

There are hundreds of periodical indexing and abstracting services. In 1963, the National Federation of Science Abstracting and Indexing Services (NFAIS) published "A Guide to the World's Abstracting and Indexing Services in Science and Technology," Washington, D.C. 1963, 183 pp. [Report No. 102]. This work lists over 1800 active indexing and abstracting services, although some deal chiefly with technical reports. Nevertheless, it is an impressive, comprehensive index, showing the great activity in this type of publication. A new edition of the **Guide** is in preparation under the joint sponsorship of NFAIS and the International Federation for Documentation (FID). It is scheduled for publication late in 1976 and will list about 2300 abstracting and indexing services.

---

[1]"CAS Publishes 5 Millionth Abstract." **Chemical and Engineering News** 49 (**50**):54, 6 Dec. 1971.

Many larger indexes rely in some measure on computers for their production. One type uses only the words of the title for its subject, with each key word appearing in its appropriate alphabetical place. For example, an article entitled "Radar Circuits for Airborne Navigation Equipment" would, in this kind of index, have its title printed five times, once under "Radar" and once under each of the other key words—"Circuits," "Airborne," "Navigation," and "Equipment." The title appears in as much of its entirety as the line permits, as shown in Figure 13-1. This is known as a KWIC (Key Word In Context) index. Obviously, the system is no better than the precision of the article title itself.

A modification of this type of index is the KWOC (Key Word Out of Context) index, in which the key word is printed out to the side of the title, which can then appear in normal word order. Sometimes the key word is printed just before a string of titles containing the same key word.

Some indexes using a KWIC or KWOC style have been programmed to permit predetermined different versions of a word to be listed together, as if they were spelled the same. For example, they might list **optical maser** with **laser,** since both mean the same thing. Or, for simplification, they might group **welds** with **welding** even though the words denote different aspects of the topic.

Another index for periodicals and some other types of literature is the **citation** index. This tool lists not only the titles and authors of the current articles but also those of previously published articles cited as references. One advantage of the citation index is that it connects an older article with more recent comments or developments relating to it.

It should be noted that many indexing and abstracting services primarily concerned with periodical articles often include a few books, certain types of reports, perhaps patents, and sometimes preprints, the early versions of what may later appear as journal articles. A notable version of this multisource type of abstracting service is **Nuclear Science Abstracts,** which is mainly for technical reports, yet includes many books and periodical articles.

Quite often the indexing services publish semimonthly or monthly issues, usually with an author index in each issue, and then publish

for hydrogenase based on an	enzymic electrode reaction. +y method	JOBIAO-0078-0443
Kinetics of the	enzymic hydrolysis of starch.=	FSPMAM-75-06-039
irion type 2 as revealed by	enzymic iodination, immunoprecipitation	VIRLAX-0067-0197
creatinine.= Kinetic	enzymic method for determining serum	CLCHAU-0021-1422
ng activation by exposure to	enzymic phosphorylating conditions.=	BBRCA9-0066-0907
lycerides+ Optimum kinetic	enzymic procedures for glucose and tri g	CLCHAU-0021-1448
yro+ Physicochemical and	enzymic properties of ATP:nucleotide p	ABCHA6-0039-1827
for electrochemically driven	enzymic reactions involving cofactors.=	BIBIAU-0017-1379
Characterization of 7-[1-(+	Enzymic synthesis of cephalosporins. II.	ABCHA6-0039-1745
in the eye of the gar and its	enzymic synthesis.= +new amino acid	TELEAY-1975-3287
atin.=	Enzymic unpacking of bull sperm chrom	BBACAQ-0403-0180
ethods for se+ Colorimetric,	enzymic, and liquid–chromatographic m	CLCHAU-0021-1427
t oxy cep halo sporanic acid	enzymically synthesized by Kluyvera cit	ABCHA6-0039-1745
and gas-sensing electrodes in	enzymology.= +c role of ion-selective,	ANALAO-0100-0609
ts pigm+ Actinomyces cyan	eofuscatus, a producer of valinomycin. I	IANBAM-1975-0694
Aldrin	ep oxidase from pea roots.=	PYTCAS-0014-1507
the kinetics of allyl bromide	ep oxidation by permaleic acid.= +on	KNKTA4-0016-0880
ygen-dependent zea xanthin	ep oxidation in isolated chloroplasts.	ABBIA4-0171-0070
3,4-diol di acetates with p+	Ep oxidation of isomeric cholest-5-ene-	BAPCAQ-0023-0557
-ene-3,4-diol mono acetat+	Ep oxidation of some isomeric cholest-5	BAPCAQ-0023-0551
trates and inhibitors in the	ep oxidation system of Pseudomonas ole	BBACAQ-0403-0245
illing by rats.= Chlor diaz	ep oxide and diazepam induced mouse k	PSYPAG-0044-0023
in rats.= Chlor diaz	ep oxide and isolation induced timidity	PSYPAG-0044-0083
II. Total synthesis of dl-crot	ep oxide.= +lo hexane derivatives. V	TELEAY-1975-3187
minor tranquilizer, chlor diaz	ep oxide.= +nt and the effect of the	PSYPAG-0044-0067
Spectral data for aliphatic	ep oxides.=	JCEAAX-0020-0445
germanium-containing	ep oxy acetates.= Silicon- and	ZOKHA4-0045-2102
estratrienes from 5,6-	ep oxy androstan-7-ols.= Formation of	JCPRB4-1975-1941
ical properties of glass fiber-	ep oxy composites.= +ess and mechan	JMTSAS-0010-1549
propylene using unsaturated	ep oxy compounds.= +fication of poly	PLMSAI-75-09-050
pmanii a+ Three new 5,10-	ep oxy germacranolides from Liatris cha	PYTCAS-0014-1803
cyclo propane ring of 6,20-	ep oxy lathyrol[1,11-di acet oxy-3,6,6,14	JCPKBH-1975-1253
ectric relaxation processes in	ep oxy resin ED-5.= +cooperative diel	VYSAAF-0017-1903
Aliphatic	ep oxy resins from general type of diol.=	CHPUA4-0025-0475
of trans-1,3-di phenyl-2,3-	ep oxy-1-propanones and their conversi	JLACBF-1975-1538
ne 21-ketones of the 20,28β-	ep oxy-18α,19βH-ursane series.= +le	CCCCAK-0040-2861
tric determination of phenyl	ephrine hydro chloride.= +Fluorome	YKKZAJ-0095-0985
e, androstenedione, dehydro	epi andro sterone, and its sulfate), estro	JCEMAZ-0041-0556
teraction between dehydro	epi andro sterone, glucose 6-phosphate d	EXPEAM-0031-1124
ation of tetra hydro furan-	epi chloro hydrin block co polymers acco	VYSAAF-0017-2120
ructure of 3β-acet oxy-6,7-	epi di thio-19-nor lanosta-5,7,9-11-tetr	JCCCAT-1975-0756
Improved synthesis of 13-	epi steroids.=	JCCCAT-1975-0748
es pugio to infection by the	epibranchial isopod parasite Probopyrus	CBPAB5-0052-0201
nt species of Staphylococcus	epidermidis.= +osition of a halo tolera	BBACAQ-0398-0464
in stratum germinativum of	epidermis in the toad Bufo bufo bufo.	CBPAB5-0052-0055
of glyceride synthesis in rat	epididymal adipose tissue.= +egulation	BIJOAK-0150-0441
tate on morphogenesis of the	epididymis in the chick.= +terone ace	CRSBAW-0169-0541
during cobalt experimental	epilepsy in the rat and its suppression b	PSYPAG-0044-0033
arbon-13 NMR spectra of	epimeric N-alkyl nor atropine derivative	JLACBF-1975-1499
acromegaly.= Plasma nor	epinephrine and epinephrine in	JCEMAZ-0041-0542
esponses by a protozoan to	epinephrine and other neurochemicals.=	SCIEAS-0190-0285
nor epinephrine and	epinephrine in acromegaly.= Plasma	JCEMAZ-0041-0542
ardi+ Depletion of heart nor	epinephrine in experimental acute myoc	EXPEAM-0031-1202
brain.= Nor	epinephrine in fetal and neonatal rabbit	EXPEAM-0031-1166
f rat liver as influenced by	epinephrine, glucagon, and hydro cortiso	BBACAQ-0404-0007
mineralizing rib and	epiphyseal cartilage.= Proteo glycans of	BBACAQ-0404-0093
tis and Escherichia coli with	episomal and chromosomal resistance.	ANTBAL-0020-0817
imperfections in germanium	epitaxial films.= Crystallographic	IVNMAW-0011-1545
allium arsenide (Al$_x$Ga$_{1-x}$As)	epitaxial films.= +eneous aluminum g	IVNMAW-0011-1700
Preparation of	epitaxial gallium nitride. II.=	MRBUAC-0010-1097
chromium-doped+ Liquid	epitaxial growth and characterization of	JJAPA5-0014-1611
ses on mica.=	Epitaxial growth of smectic ordered pha	JOPQAG-0036-1029
tantalum during the the+	Epitaxial growth of tantalum deposits on	JCRGAE-0029-0367
wth and characterization of	epitaxial layers of lead tin telluride (Pb$_1$	JCRGAE-0029-0241
of undoped gallium nitride	epitaxial layers.= Cathodoluminescence	FTPPA4-0009-1772
-ups, threading+ Defects in	epitaxial multilayers. II. Dislocation pile	JCRGAE-0029-0273
V) thin films by liquid phase	epitaxial techniques.= +hium niobate(	JCRGAE-0029-0289
bstrate in LPE [liquid phase	epitaxial] saturation process.= +ide su	JJAPA5-0014-1613
films prepared by the liquid	epitaxy method.= +phide (Ga As$_{1-x}$P$_x$)	IUZFAU-19-04-077

Figure 13-1. Portion of a page from **Chemical Titles**, illustrating a KWIC index. Copyright American Chemical Society; reproduced with permission.

bound annual cumulative volumes with annual author indexes. There are many exceptions to this pattern, however.

SELECTED EXAMPLES

General—Science and Engineering

Applied Science and Technology Index. New York, H. W. Wilson, 1913- [Monthly; quarterly and annual cumulations]

This work covers all aspects of engineering and other applied sciences, although the emphasis in the latter area is on physical science. Each article may be listed under several subject headings. The work indexes over 200 periodicals, usually including all articles and features, such as biographical sketches, statistics, and so on. It is restricted to English language journals, mostly of the U.S., and, despite its lack of abstracts and author index, it is a very useful service. It is found in many public libraries and most college libraries, even those where engineering is not a major interest.

Engineering Index. New York: Engineering Index, Inc., 1884- [Monthly; annual cumulation]

This is probably the best general index to engineering literature, dealing, as it does, with all aspects of the subject, with indicative abstracts, and having a wide journal coverage (over 2,200 publications). It cites periodicals as well as proceedings and special publications of such groups as engineering societies, scientific and technical associations, universities, laboratories, and government agencies. All issues include an author index and have many cross-references. In 1974, over 85,000 items were indexed. The index is also available on microfilm and in a magnetic tape version called COMPENDEX (see Chapter 28). Figure 13-2 shows a sample of **Engineering Index.**

Science Citation Index. Philadelphia: Institute for Scientific Information, 1961- [Quarterly; annual cumulations]

This work indexes the sources cited in current articles in over 2,400 scientific and technical periodicals. Among other features, in later years it allows one to trace the reference use of a given older article. It has separate personal author and corporate author indexes, and a companion publication, **Permuterm Subject Index,** provides a permuted index to the titles of all current articles. The service is also available as a weekly magnetic tape service, for computer searching (see Chapter 28). One five-year cumulation—for 1965-69—was published and one is available for **1970-1974.**

077662

TITANIUM ALUMINUM MOLYBDENUM ALLOYS—*Contd.*
Mechanical Properties See TITANIUM METALLOGRAPHY—Phase Diagrams.

TITANIUM ALUMINUM VANADIUM ALLOYS

Mechanical Properties

077662 INCIDENCES DE L'ETAT MICROSTRUCTURAL SUR LA DUCTILITE ET LE COMPORTEMENT A LA RUPTURE DE L'ALLIAGE DE TITANE TA6V. [Effect of the Structural State of the TA6V Alloy on Toughness and Fracture Behavior]. The differential structural states of the Ti-6Al-4V alloy were obtained by different thermal and thermomechanical treatments. The best mechanical properties were the result of thermomechanical treatment in the alpha-plus-beta range. Thermal treatment in the beta state resulted in low ductility. After either treatment the energy of rupture values were very similar between −196 and +200 C, and fracture was ductile. Impact strength, however, was affected by temperature, varying from 2 daj/cm² at −196 C to 7 daj/cm² at +200 C. In French.

Hadj Sassi, B. Cent d'Etud de Chim Metall, Vitry-sur-Seine, Fr; Quoix, Ph.; Lehr, P. *Mem Sci Rev Metall* v 72 n 4 Apr 1975 p 341-355.

Sheet and Strip

077663 INFLUENCE OF PRIOR TEXTURE ON THE COLD ROLLED TEXTURE OF Ti-6Al-4V. In most published studies of cold rolled textures of titanium and titanium alloy sheet, the amount of reduction is large, and it is assumed that any effects of the interphase boundaries in the alpha-plus-beta Ti-6Al-4V alloy lead to the conclusion that the interface phase is neither titanium hydride nor simply a region of high dislocation density. It is a distinct layer which develops either as hexagonal α-phase having a different orientation than the primary α into which it grows, or as an fcc phase, depending upon heat treatment. Such complex interphase boundary structures seem to be common in all titanium alloys containing alpha/beta boundaries which have formed at about 675-875 C.

Rhodes, C.G. Rockwell Int, Thousand Oaks, Calif; Williams, J.C. *Metall Trans A* v 6A n 8 Aug 1975 p 1670-1671.

Phase Diagrams

077666 THERMAL EQUILIBRIA AND MECHANICAL STABILITY OF Ti₃Al PHASE IN Ti-Mo-Al ALLOYS. The phase diagram of the isopleth section of the Ti-Al-7Mo (at.%) system was improved and expanded to include alloys with up to 25 at.% Al. The mechanical and thermal stability of alloys aged in the two-phase region, β+Ti₃Al, was correlated with the microstructure. X-ray rocking-curve studies of the polycrystalline specimens showed that after 2% deformation of a Ti-16Al-7Mo alloy the beta matrix became preferentially plastically deformed, while the Ti₃Al particles functioned as hard particles undergoing little lattice distortions.

Hamajima, T. Rutgers Univ, New Brunswick, NJ; Weissmann, S. *Metall Trans A* v 6A n 8 Aug 1975 p 1535-1539.

Precipitation See TITANIUM VANADIUM ALLOYS—Mechanical Properties.

Specimen Preparation

1767-1774.

TITANIUM MOLYBDENUM ALLOYS

Deformation

077670 PLASTIC DEFORMATION AND DISLOCATION KINETICS IN α, (α+β) and β-TITANIUM ALLOYS. The temperature and strain-rate dependence of the flow stress of an (α+β) Ti-6.5 at.% Mo alloy was investigated at 77 to 823 K and strain rates of 0.6 to 1.5×10⁻⁴ /sec. The results of a thermal activation analysis are compared with those of previous work on close-packed hexagonal α titanium, and the β bcc Ti-15.2 at.% Mo alloy with different interstitial contents. The activation parameters derived for the three types of alloy were in good agreement. It is concluded that thermally activated overcoming of interstitial solute atoms is the rate determining mechanism in all these alloys. Molybdenum and microstructure mostly affect the athermal component of the flow stress. 40 refs.

Zeyfang, Rolf R. Univ of Ky, Lexington; Conrad, Hans. *Z Metallkd* v 66 n 7 Jul 1975 p 422-427.

Mechanical Properties See POWDER METALLURGY—Sintering.

TITANIUM OXIDES See PAPERMAKING—Coloring.

TITANIUM VANADIUM ALLOYS

Deformation See TITANIUM METALLOGRAPHY—Transformations.

Mechanical Properties

Figure 13-2. A portion of a representative page from Engineering Index. Reproduced with permission of Engineering Index, Inc.

Aeronautics and Astronautics

International Aerospace Abstracts. New York: American Institute of Aeronautics and Astronautics, 1961-   [Semimonthly]

Published for the Institute and for NASA, this work covers aeronautics and space science and technology. It indexes periodicals, books, and conference papers and has indicative abstracts. Each issue has author, subject, contract number, and report index number indexes. There are semiannual and annual cumulative indexes for subjects as well as for the types of items found in the indexes in each issue (author and so forth). Over 1,200 journals are scanned regularly for items to index. It is a companion volume to **Scientific and Technical Aerospace Reports.**

Biology

Biological Abstracts. Philadelphia: BioSciences Information Service of Biological Abstracts, 1927-   [Semimonthly]

This abstracts periodicals covering a wide range of the life sciences, including zoology, botany, biochemistry, physiology, etc. In each issue there are several indexes, including authors and subjects; indexes cumulate semiannually. Also available on microfilm and magnetic tape (see Chapter 28). There is a companion publication, **BioResearch Index**, which covers reports, symposia, trade journals, etc.

Business

See **Business Periodicals Index**, Chapter 25.

Chemical Engineering and Chemistry

Chemical Abstracts. Columbus, Ohio: American Chemical Society, 1907-   [Weekly]

Probably the best abstracting service in the scientific and technical world, CA indexes, on a selective basis, over 12,000 journals, patents for 26 countries, new books, conference proceedings, and technical reports. The abstracts are very informative, giving explicit details such as names of all compounds, temperatures, quantities, and equipment used. It has weekly and semiannual author, subject, and patent indexes. Decennial (1907-1956) and quinquennial (since 1956) cumulative indexes are available. Some indexes include cumulative formula indexes, an index to ring systems, a registry (chemical), and number index. CA is available in microfilm as well as on magnetic tape (CHEMCON) (see Chapter 28). Around one-third of a million items, grouped into 80 subject categories, are indexed each year.

CA covers a wide range of subjects besides pure chemistry, such as chemical engineering, metallurgy, solid state studies, and so on. It is also available in five sections for those who do not wish the entire contents. Other specialized indexes are available. See Figure 2 for a sample page.

Civil Engineering

ASCE Publications Abstracts. New York: American Society of Civil Engineers, 1966-   [Bimonthly]

This publication contains indicative abstracts of all papers appearing in all the ASCE journals and in **Civil Engineering.** It covers the fields of engineering mechanics, hydraulics, irrigation, sanitary engineering, soil mechanics and foundations, surveying and mapping, urban planning, and general civil engineering. The tables of contents are reproduced, with a separate subject index and separate citations (in 3 X 5 inch format), for those who wish to clip and file. It also has author indexes but no cumulative indexes.

Building Science Abstracts. Garston, Watford, Engl.: Building Research Station, 1928-   [Monthly]

Indicative abstracts are given for selected periodical articles. The work considers construction materials as well as techniques and operations and has annual author and subject indexes.

Highway Research Abstracts. Washington, D.C.: National Research Council, Highway Research Board, 1931-   [Monthly; annual subject index]

This is a very selective index for reports and periodical articles on all aspects of highways—technical, safety, economic, and so on. It has brief annotations.

Computers and Data Processing

Computer and Control Abstracts. London: Institution of Electrical Engineers, 1966-   [Monthly]

Jointly published with the Institute of Electrical and Electronics Engineers and issued as Part C of **Science Abstracts,** this abstract service for periodical literature in the fields of computers and controls gives broad subject coverage. Items have indicative abstracts and are arranged by a classification plan. Each issue has separate author, patent, conference paper, and book indexes, which cumulate semiannually. A cumulative author and subject index is available for the period 1969-1972. It is also available on magnetic tape. (See Chapter 28.)

Computing Reviews. New York: Association for Computing Machinery, 1960- [Monthly]

This abstract of books, theses, and periodical articles gives detailed annotations, approaching the informative style. Dealing with programming, equipment design and construction, applications, social and economic aspects, and so forth, it has an annual author index and a separately issued permuted subject index.

Electrical Engineering

Electrical and Electronics Abstracts. London: Institution of Electrical Engineers, 1898- [Monthly]

This joint publication of the British society named above and the Institute of Electrical and Electronics Engineers is issued as Series B of **Science Abstracts.**

It has rather informative abstracts, and each issue contains indexes for authors, bibliographies (either those items written specifically as bibliographies or those containing over 50 citations), books (to emphasize papers from such sources), patents (by patent number), and reports (by report number). Semiannual cumulative indexes are issued, as well as a cumulative author and subject index for 1969-1972. It scans over 2,000 journals for items and indexes 100 completely. The work is closely related to **Physics Abstracts** (Series A of **Science Abstracts**) and **Computer and Control Abstracts** (Series C of **Science Abstracts**). It is also available on magnetic tape. (See Chapter 28.)

Electronics

See **Electrical Engineering.**

Energy

Energy Index. New York: Environment Information Center, 1973- [Annual]

This work is an index of periodical articles, laws, documents, books, films, and statistics. It has author and subject indexes.

Environment

Environment Index. New York: Environment Information Center, 1971- [Annual]

This work presents citations to some 75,000 items from periodicals, and also lists important books, conferences, legislation, and films. It has indexes by subject, author, and geography, and includes a directory of federal and state environmental officials.

Industrial Engineering

International Abstracts in Operations Research. Baltimore, Md.: International Federation of Operational Research Societies, 1961- [Bimonthly]

> This index is restricted to operations research but covers all aspects of the subject, including applications. It has indicative abstracts, some reprinted from other sources. Author and subject indexes are in each issue as well as in annual cumulations. Over 75 journals are regularly indexed.

Quality Control and Applied Statistics. Whippany, N.J.: Executive Science Institute, 1956- [Monthly]

> This work offers very detailed, informative abstracts of selected periodical articles. Issued in loose-leaf form, it has annual author and subject indexes. It is closely related to a companion publication, same source, entitled **Operations Research Management Science**, which began in 1961.

Mathematics

Mathematical Reviews. Providence, R.I.: American Mathematical Society, 1940- [Monthly]

> This is an annotated index to the world's mathematical journals plus a certain number of books, theses, and so on. Each issue has an author index, which cumulates semiannually. There are also cumulative author indexes that cover several years, as well as semiannual subject indexes.

Mechanical Engineering

Applied Mechanics Reviews. New York: American Society of Mechanical Engineers, 1948- [Monthly]

> The major topics covered are mechanics of solids, mechanics of fluids, automatic control, heat, rational mechanics and mathematical methods, and combined fields. It has indicative abstracts and author indexes in each issue and there are annual cumulative author and subject indexes, although the latter is by key words and rather inadequate.

Metals and Metal Working

Metals Abstracts. London: Institute of Metals, 1968-   [Monthly]

Jointly published by the Institute of Metals and the American Society for Metals, this work replaces their former publications, **Metallurgical Abstracts** and **Review of Metal Literature**, respectively. Each issue has indicative abstracts and an author index, and is a companion volume to **Metals Abstracts Index**, a monthly publication containing a subject index (plus the same author index). Cumulative author and subject indexes are issued. Around 25,000 abstracts are indexed per year and over 1,200 journals are covered. It is also available on magnetic tapes. (See Chapter 28.)

Mining and Minerals

IMM Abstracts. London: Institution of Mining And Metallurgy, 1950-   [Bimonthly]

The index deals with mining, mineral dressing and economic geology. It is arranged by a subject classification system, and all items have annotations that are often more informative than indicative in style.

Nuclear Engineering

See **INIS (Atomindix)** and **Nuclear Science Abstracts**, Chapter 15.

Ocean Engineering and Oceanography

Oceanic Abstracts. La Jolla, Calif.: Pollution Abstracts, 1963-
[Bimonthly]

This publication covers all aspects of the oceans, including the phases of interest to biology, geology, fishing, meteorology, pollution, and so forth. Sometimes it has a brief abstract and it consistently has at least a list of key words for each item indexed. Each issue has an author and a key word index, which cumulate annually. It was formerly titled **Oceanic Index** and **Oceanic Citation Journal.**

Physics

Physics Abstracts. London: Institution of Electrical Engineers, 1898-
[Semimonthly]

Published as Part A of **Science Abstracts,** this work thoroughly covers all aspects of physics as well as astronomy, with all items annotated. Each issue has a separate author index. Semiannual cumulative indexes are issued, as well as a cumulative author and subject index for 1969-1972. Current awareness service and magnetic tapes are also available. (See Chapter 28.)

Pollution

Air Pollution Abstracts. Washington, D.C.: U.S. Environmental Protection Agency, 1970- [Monthly]

This index to literature recently acquired by the Air Pollution Technical Information Center includes books, periodical articles, technical reports, and the like. The items have indicative abstracts and are arranged by 14 subject categories. Each issue has its own author and subject indexes, which cumulate semiannually.

Pollution Abstracts. La Jolla, Calif.: Pollution Abstracts, Inc., 1970- [Bimonthly]

An index to books, periodical articles, and conference papers, this work is arranged by broad categories, with brief annotations. Each issue has its own author and permuted title indexes. A two-year cumulative index is available, as well as annual indexes.

Water Pollution Abstracts. London: H M. Stationery Office, 1927- [Monthly]

This index to periodical literature has an annual author and subject index. Issues contain rather informative abstracts.

CHAPTER 14

Technical Reports

Periodical articles are usually much more up to date than books. Similarly, technical reports are usually much more up to date than periodical articles, especially those in scholarly journals. These reports are usually from a company or government agency, and may range from a few typewritten sheets to a full-size book. They are almost always from noncommercial publishers—agencies that do not have as their primary goal the publication and sale of these reports. For example, when NASA issues a report, this is obviously not its primary goal. The report merely summarizes the results of research. Thus, technical reports can be properly called unpublished information. These reports are also usually part of a numbered series.

In addition to being more timely, technical reports can be more specific than periodical articles, since the writers have even less cause than periodical editors to seek out a specific audience. They are usually aimed at those readers having a good background in the subject, and rarely take a tutorial approach.

Some disadvantages of technical reports are that they are often not as well edited as published literature, they are more difficult for the nonqualified person to comprehend, and they are more difficult to locate. Sometimes it is difficult even to learn of their existence.

Two types of security regulations often affect a large percentage of the existing technical reports—military and commercial. Military, or national, security regulations require that a person be cleared by the government before being allowed to use classified reports and, further, that there be a "need to know" the information. The site at which the information is to be used must also have been cleared.

Reports that are commercially classified, or "company confidential," are restricted to protect the success of the company.

Every report has it own serial code and number, supplementing the listing of the usual bibliographic information such as author, title, date, and so forth. Of all the elements used to describe reports bibliographically, the report series code and report number are the most important. A useful directory of codes is cited below.

Because of the great and increasing number of technical reports, many are issued on microfilm to save space and reduce the costs of publication. The favored type of microform for reports is the microfiche, which is simply a piece of microfilm about 4 x 6 inches on which there are up to 90 or more individual frames, each of a page from the report. Microfilm reading equipment—which enlarges the image approximately 25 times—is required. Many libraries and information centers also have reader-printers, which enable one both to read the fiche and to make full-size enlargements to keep. Some technical reports were issued on the earlier 3 x 5 inch microcard, but it was hard to read and to enlarge. Practically nothing new is being issued on microcards.

SELECTED EXAMPLE

Dictionary of Report Series Codes, 2d ed. Edited by Lois E. Godfrey and Helen F. Redman. New York: Special Libraries Association, 1973. 643 pp.

> This index to codes used to identify technical report services is arranged by code and identifying source, and also by source with the accompanying code. Very useful in locating reports.

CHAPTER 15

Technical Report Indexing and Abstracting Services

Indexing services for periodical literature are important, but indexes for technical reports are far more so because it is much more difficult to learn of the existence of these reports and often still more diffciult to obtain them once their existence is known.

A good report indexing service, besides having the usual indexes, such as subject and author, should also list the names of agencies or groups preparing the reports, the report numbers, and the government contract numbers under which reports are prepared. Sometimes the only information one has is one of these items. Therefore, an index should allow for a search in each of these categories, as well as in the traditional author-subject format.

Some of the major report indexes now include all these elements as a matter of course, in an effort to make it easier for reports to be located and obtained. A few agencies—and most of them are U.S. government-affiliated, usually with the agencies that create the reports—provide extra services for organizations that have contracts with them, such as free copies of the agencies' own indexing and abstracting publications, free copies of reports, and free searching and reference services. The variability of government budget restrictions causes variations in the extent of free services available to contractors, so it is not possible to state concretely what will be available at any one time. As for service to the public in general, some agencies have sales offices from which most unclassified documents may be obtained or ordered by mail.

One notable example of this spirit of promoting the availability of reports is found in the Department of Commerce, which has had a congressionally imposed responsibility since World War II to make unclassified government reports of general interest to the public available for sale. There have been many names for the section or office having this responsibility, the latest dating from 1970. It is now known as the National Technical Information Service, and it is charged with indexing and promoting the sale of reports from the Department of Defense, the Department of Commerce, and other agencies such as the Federal Aviation Agency. Its series are generally recognized by the letters **PB** listed before the document number (as **PB** 14111), the letters dating back to its original designation as the Publication Board.

The military counterpart to this group has also been in existence since the end of World War II and has gone through various name changes and reorganizations. It is now known as the Defense Documentation Center, and its reports usually use the prefix **AD**, where again the letters stand for a previous agency name (Armed Services Technical Information Agency Documents, or ASTIA Document). Its report abstracting service, **Technical Abstract Bulletin** (or **TAB**), is listed only as a means of checking older literature, since in 1967 it became a classified document itself, thus taking it beyond the scope of this book. Needless to say, there are various indexing services for classified military documents with which engineers working in such areas should become familiar.

The Energy Research and Development Administration (ERDA) index, **Nuclear Science Abstracts** (which indexes not only technical reports but also books and journals), is very well prepared. Its report index indicates where a given report can be obtained and includes the very useful feature of citing books or periodical articles based on the report. This is an extremely handy feature, since it is most discouraging to search for a report only to find later that its equivalent was already available as a periodical article or book based on the original report. Effective July 1976 this publication was replaced by **INIS Atomindex** (see citation below).

The index issued by NASA for its reports, **STAR** (Scientific and Technical Aerospace Reports), is equally well prepared, although it lacks the special feature just mentioned for alternate location of a report.

Not all report series of national import are included in the indexes mentioned above; some are indexed in the Superintendent of Documents' index, **Monthly Catalog for Government Publications**. These are usually congressional documents (hearings, bills, and so on) as well as reports from the social services and the financial and business end of government, rather than from technical agencies. Nevertheless, they are often of great importance.

SELECTED EXAMPLES

ERDA Energy Research Abstracts. Oak Ridge, Tenn.: U.S. Energy Research and Development Administration, 1975-   [Monthly]

> This work indexes nonnuclear energy reports, patents, journal articles, conference papers, theses, and books originated by ERDA and its contractors, plus some foreign sources. It covers coal, petroleum, solar energy, geothermal energy, and related topics, such as materials. It also has indexes for agencies, authors, subjects, and report numbers.

Government Reports Announcements and Index. Springfield, Va.: U.S. National Technical Information Service, 1946-   [Semimonthly]

> This index covers a wide range of report literature, as it includes technical, scientific, business, and related documents prepared by or for most government agencies, including the Department of Defense. It includes some NASA and ERDA reports, and continues **U.S. Government Research and Development Reports** (and other titles), which it supersedes. It has indexes for authors, agencies, subjects, report numbers, and contract numbers in each issue, which cumulate annually. 24 different weekly newsletters **(Weekly Government Abstracts)** provide current awareness data. It is also available on magnetic tape. (See Chapter 28.)

INIS Atomindex. Vienna: International Atomic Energy Agency, 1970-   [Semimonthly]

> This publication, a product of the International Nuclear Information System (INIS), is devoted to the fields of nuclear energy and its applications. It provides abstracts for books, patents, technical reports and periodical articles. Each issue has author, subject, report number and patent number indexes, which cumulate semiannually. Supersedes **Atomindex** (1959-1970). In July 1976 superseded **Nuclear Science Abstracts** for coverage of ERDA reports in this field.

Monthly Catalog of United States Government Publications. Washington, D.C.: Superintendent of Documents, 1895-   [Monthly]

This index of recently published reports, hearings, documents, serials, pamphlets, and related materials, all issued by federal government agencies, is arranged by issuing source with monthly author, title, and subject indexes, which cumulate annually. No annotations are given. One issue (February) lists all the serials and periodicals issued by the government.

Nuclear Science Abstracts. Oak Ridge, Tenn.: U.S. Energy Research and Development Administration, 1948-1976. [Semimonthly]

The issues of this abstracting service cover nuclear science and technology on an international scope, including not only reports but also books, journals, patents, and so forth. All items are annotated. Each issue has separate indexes for subjects, personal authors, agency sources (corporate authors), and report numbers. Indexes cumulate quarterly and annually; some cover several years. (ERDA was formerly known as the Atomic Energy Commission.) Replaced by **INIS Atomindex**, beginning July 1976.

Scientific and Technical Aerospace Reports. College Park Md.: U.S. National Aeronautics and Space Administration, 1963- [Semimonthly]

This abstracting service, designed to cover the field of aeronautics and astronautics, both scientific and engineering aspects, is restricted to report literature (for published literature, see **International Aerospace Abstracts**). Each issue has separate indexes for personal authors, subjects, agency sources (corporate authors), contract numbers, and report numbers, which cumulate semiannually and annually.

Selected RAND Abstracts. Santa Monica, Calif.: RAND Corporation, 1946-   [Quarterly]

An annotated index to the unclassified RAND reports, which cover not only technical and scientific topics but also economic and political subjects, each issue also has separate author and subject indexes, which cumulate annually. There is a cumulative index for 1946-1962.

Technical Abstract Bullettin. Springfield, Va.: U.S. Defense Documentation Service, 1953-1967.

TAB was formerly the source of information about classified and unclassified Department of Defense-sponsored reports (although it itself was not classified). Now unclassified DOD reports are indexed in **Government Reports Announcements & Index.** TAB became a classified publication in 1967.

Part IV
Other Sources of Information

This section describes technical information that will probably be less familiar to most readers than the material in the previous sections. The types of printed formats introduced here include patents, tabular data, newspapers, maps and graphic forms, dissertations, manufacturers' catalogs, and specifications or standards.

This section also describes alternate sources to printed or published works—including personal sources, government agencies, professional societies, and educational institutions—as well as special services and materials, such as current awareness services, the use of machine-readable data files, and retrospective searching for information.

CHAPTER 16

Conferences and Symposia

One of the major ways of keeping up with a given field is by attending the conferences and meetings sponsored by technical groups. Many authors wait for just this kind of audience to make public their most important finds. Information given out at meetings can thus precede regular publication by several months or more.

Unfortunately, most professionals cannot afford to attend all the meetings they might want to. In such cases, there are two reasonably equal choices open to those interested in knowing what's going on—either to obtain the preprints, or author-prepared summaries, usually available upon application to the conference sponsors, or to write directly to the authors for copies of what was distributed by them at the conference.

A third solution is to await publication of the official proceedings of a conference. Here, the word "proceedings" denotes the collections of all the papers given, usually edited for printed presentation. Publication may be at any time from immediately after the meeting (usually possible only when authors are restricted to a 300-500 word abstract of their papers) to periods of up to three or four years later. The specificity of most papers tends to make them more useful for the experienced engineer than for the neophyte. The subject indexing, often sketchily done, adds to the difficulties, as does the uneven coverage by the indexing and abstracting services, which makes the retrieval of information from conferences highly uncertain in the first place. The varied format of proceedings also is of no help, since they can appear as books, periodical articles, supplements to journals, separate publications, or parts of technical report series. Thus this category of information cuts across many boundary lines.

This chapter does not attempt to indicate all the major technical meetings for which printed proceedings are regularly issued; such a list would be voluminous. Also not indicated are many valuable one-shot meetings for which proceedings are prepared.

Trade and professional journals usually carry calendars of forthcoming meetings that are useful in keeping engineers aware of conferences worth attending. (Practically every professional and technical society sponsors one or more each year.) In addition, two references—**Directory of Published Proceedings** and **Proceedings in Print**—are especially helpful when one wants to know about past meetings—whether proceedings were published, and if so how and where to find them.

SELECTED EXAMPLES

Current Programs. Chestnut Hills, Mass.: World Meetings Information Center, 1973- [Monthly]

> This work lists titles of papers given at national and international scientific and technical meetings. There are quarterly indexes, which cumulate annually, for subjects, authors and dates of meetings. Lists thousands of papers, as well as sponsors of meetings.

Directory of Published Proceedings. Series SEMT. Harrison, N.Y.: Inter Dok Corp., 1964- [Monthly; annual cumulations]

> This chronological history of published proceedings and preprints for conferences, congresses, symposia, meetings, seminars, and summer schools is international in scope. The series is devoted to science, engineering, medicine, and technology. It gives information about the title and location of the meeting plus full details about the availability of printed proceedings. This is a valuable tool for an elusive set of publications. Indexes, editors, locations, subjects, and sponsors, with annual cumulations.

Irregular Serials and Annuals. 2d ed. New York: Bowker, 1972. 850 pp.

> This index to some 18,000 serials, annual publications, symposium proceedings, supplements, and so on, is restricted to publications published not more often than once a year.

CONFERENCE AND SYMPOSIA

Proceedings in Print. Mattapan, Mass.: 1964-   [Bimonthly]

An index for conference proceedings in all disciplines and all languages, this volume uses a keyword index of conference titles plus indexes for sponsors of meetings. Those meetings for which proceedings will not be published are also listed. Indexes cumulate annually.

Scientific Meetings. New York: Special Libraries Association, 1957- [Quarterly]

This publication lists meetings of interest to scientists and engineers, with indexes by name of sponsoring group, by date, and by subject. It also gives addresses of sponsors.

CHAPTER 17

Specifications and Standards

The terms **specifications** and **standards** are sometimes used almost synonymously. Both refer to rather detailed instructions about how something is to be manufactured, managed, designed, or otherwise handled. Usually, however, **specifications** are limited to a certain application, such as the specifications for making one particular bridge. Conversely, **standards** imply a more universal application, such as the strength requirements for all highway bridges in a particular state.

Our world would be infinitely more complicated if standards did not exist. Even buying a light bulb would become a complex matter if, for each lamp, we had to determine type of socket threads, diameter of the socket, house voltage, and so forth. Fortunately, standards have existed for years for certain basic types of electrical equipment so that we need only know the wattage of a bulb to find one to replace it. Such convenience did not just happen. Somewhere along the line manufacturers and electricians and perhaps architects had to sit down together and decide upon dimensions, electrical features, and so on. These standards were then approved by the industries and trades involved and accepted for general use by manufacturers.

There are two main sources of specifications and standards—government and nongovernment, with government being by far the more prolific. Government sources can then be further subdivided into military and nonmilitary, with military being the more prolific. Formerly, each military service issued its own specifications for all its equipment, but the overlap was tremendous and military standarization activities were merged. In addition, the General Services Administration (GSA) in Washington was charged with preparing standards for common items, such as light bulbs, canned food, and tools. These GSA publications are called federal specifications.

Each of these two government standards agencies issues indexes to its standards, usually annually, with monthly supplements.

Of the nongovernment sources, one of the larger publishers of standards is the multidisciplinary agency known as the American Society for Testing and Materials (ASTM). It issues some 30 bound volumes per year on a score or more topics, such as petroleum products, concrete, wood, and metals.

Agencies representing different trades and professional groups also issue standards. The national coordinating agency for them all is the American National Standards Institute (ANSI). Its committees prepare standards on every topic from the design of automobile brakes to test methods for nuclear reactors, although it often accepts the standard from a group such as the ASTM and renumbers it with an ANSI number.

Another nongovernment sector that issues standards is private industry. Large corporations usually have self-determined standards for their manufacturing and design groups, although the public rarely sees them.

Obviously, some areas of science, such as mathematics and theoretical physics, have little need for standards, but most engineers will need them—especially those who work on government contracts. And considering the present importance of the consumer-government influence on the safety and effectiveness of mass-produced items, it is likely that the influence of standards and specifications on engineering practices will grow considerably. Every manufacturer will be increasingly concerned with them, whether they are self- or government-imposed.

SELECTED EXAMPLES

American National Standards Institute. **Catalog.** New York: The Institute, 1948(?)- [Annual]

The ANSI **Catalog** provides a listing of available standards, arranged by subject categories. Supplements are issued irregularly.

American Society for Testing and Materials. **ASTM Standards.** Philadelphia, Pa.: 1939-   [Annual]

The compilation of standards emphasizes tests for a wide range of materials. Over 30 volumes are issued each year, including an index volume.

Struglia, Erasmus J. **Standards and Specifications Information Sources.** Detroit: Gale Research Company, 1965. 187 pp. (Management Information Guide No. 6)

Sources cited for this type of literature include general ones, government, associations and societies, and international. The work lists periodicals and bibliographies of interest and has author-title and subject indexes. The items are annotated.

U.S. Department of Defense. **Index of Specifications and Standards.** Washington, D.C.: 1951-   [Annual]

Issued with cumulative bimonthly supplements, this publication lists all unclassified specifications and standards adopted by the Department of Defense. One volume is arranged alphabetically by title, one by a classification system, and one by the standard numbers. Availability is indicated, as well as number and date of latest edition, and so on.

U.S. General Services Administration. **Index of Federal Specifications and Standards.** Washington, D.C.: Superintendent of Documents, 1952-   [Annual; monthly supplements]

This work lists nonmilitary specifications accepted for federal use. It is arranged by subject classification, by number, and alphabetically by title, and includes price, date, edition number, and so forth.

CHAPTER 18

Manufacturers' Catalogs and Directories of Manufacturers

Trade catalogs, or manufacturers' catalogs, are sometimes indispensable—such as when the design of large projects requires that the engineer know typical dimensions or performance figures for smaller units or pieces of equipment. Manufacturers' catalogs usually give this sort of information. Because these catalogs are occasionally inaccurate, however, and often difficult to get, it may be helpful to use that adjunct to some directories of manufacturers—the listing of names and locations of the manufacturers' local sales representatives.

Because of the difficulty of keeping a collection of catalogs complete and up to date, some libraries have turned to commercially prepared microfilm services. These sets, sold by subscription, are automatically—and frequently—updated and come equipped with printed indexes arranged by company and by product. One type is arranged by vendor, with a product catalog to supplement it, while another is arranged by type of product, with an index arranged by manufacturer.

Still another source of the catalogs themselves is the bound sets, from participating companies, that are printed and distributed by a publisher. One of the chief examples of this is the service established primarily for architectural materials, machine tools, and certain manufacturing industries (see **Sweet's Catalog File**).

This chapter also includes a listing of general and specific directories of manufacturers.

SELECTED EXAMPLES

General

MacRae's Blue Book. Hinsdale, Ill.: 1910-   [Annual]

This annual directory of manufacturers is a multivolumed set, with most space given to a classified-products directory. One volume gives 'manufacturers' addresses, and a reent addition is another volume which contains several thousand pages of catalogs from many manufacturers.

Thomas Register of American Manufacturers and Thomas Register Catalog File. New York: Thomas Publishing, 1905-   [Annual]

This comprehensive directory of thousands of American manufacturing companies has a product arrangement of manufacturers as well as a separate list of company addresses and a list of brand names and trademarks. Also, separate volumes now present reproductions of companies' trade catalogs. The most recent set of this annual publication runs to 11 volumes.

VSMF Design Engineering System. Englewood, Colo.: Information Handling Services. [Monthly]

This is a compilation on microfilm cartridges of vendors' catalogs and product specification sheets. It covers thousands of products, and has a monthly updating service. It is arranged by type of product. (Other available services also index standards and data compilations in like manner.)

Zimmerman, O. T., and I. Lavine, eds. **Handbook of Industrial Trade Names.** Dover, N.H.: Industrial Research Service, 1953-1965. 1 vol. and 4 supplements.

An alphabetical listing of thousands of trade names, giving their characteristics, this handbook also has an index by type of use and a directory of manufacturers involved.

Specific Directories

Best's Safety Directory: Safety, Industrial Hygiene, Security. Oldwick, N.J.: A. M. Best Co., 1946-   [Biennial]

This is essentially a directory of safety and pollution-control equipment manufacturers of all descriptions. The categories include fire protection, machinery guards, industrial hygiene, transportation safety, plant maintenance, air and water pollution controls, and so forth.

MANUFACTURERS' CATALOGS

Chemical Engineering Catalog: The Process Industries' Catalog. New York: Van Nostrand Reinhold, 1916-   [Annual]

This collection of manufacturers' catalogs on processing equipment and related materials is arranged by company, with indexes by product and by trade name.

Chemical Materials Catalog. New York: Van Nostrand Reinhold, 1949-   [Annual]

This compilation of catalogs from manufacturers of chemicals is arranged by company, with supplementary indexes by product and by trade name.

Computer Directory and Buyers' Guide. Newtonville, Mass.: Berkeley Enterprises, 1974-   [Annual]

Issued each June, by the publishers of **Computers and People** (periodical), this work contains a wealth of information covering data processing equipment, its uses, and its users. It includes a buyers' guide to products and services, a roster of college computer facilities, a who's who in data processing, a roster of computer associations, and special studies involving characteristics of digital computers, a listing of over 2,000 computer applications, and so on.

Electronic Buyers' Buide. New York: McGraw-Hill, 1946-   [Annual]

This publication contains a detailed list of manufacturers' addresses, then a product arrangement of vendors.

Jane's World Mining—Who Owns Whom,—The World Companion to Mining Investment. New York: McGraw-Hill, 1970. Unpaged.

This book, aimed at the mining investor, lists more than 2,300 companies alphabetically, shows ownership and subsidiary interests, and includes joint ventures.

Keystone Coal Industry Manual. New York: McGraw-Hill. 1918- [Annual]

This publication consists of a combination of technical articles on the mining and sale of coal as well as a directory of United States and Canadian mining companies, equipment manufacturers, utilities, coal buyers, coal seams, and so on.

Metal Finishing Guidebook Directory. Westwood, N.J.: Metals and Plastics Publications, 1930-   [Annual]

Included as an issue for subscribers to **Metal Finishing** periodical, the directory serves as a handbook, covering such topics as mechanical and chemical surface preparation, plating solutions, testing, organic finishing, and the like. It includes directories of manufacturers, technical societies, and so forth.

Mining Year Book. London: Mining Year Book Ltd., 1887- [Annual]

This work consists of an alphabetical listing of mining companies on an international basis. It gives officers, types of business, financial status, and so on.

Sweet's Catalog File. New York: McGraw-Hill Information Systems Company, 1914- [Annual]

These compilations of manufacturers' catalogs are centered around product lines such as machine tools, architectural materials, and plant engineering.

CHAPTER 19

Tables, Statistics, and Data Compilations

Much of the information an engineer needs is in tabular, statistical, or similar compact form (data compilations), as distinct from more descriptive literature.

Most **tables** are directly concerned with mathematics or with engineering and scientific formulas—one of the best sources for these is **An Index of Mathematical Tables**, a classic work in its field, which covers both common and esoteric tables. An adjunct to this book is the periodical **Mathematics of Computation** (formerly titled **Mathematical Tables and Other Aids to Computation**), which contains many special mathematical tables. Together these constitute an extremely useful set. For a general, all-around valuable source of data, the **Handbook of Chemistry and Physics** would rate near the top of the list.

Occasionally tables are found in periodical or technical report formats, but the most useful ones seem to find their way into book form. One exception to this "rule" is typified by the U.S. Bureau of the Census decision in recent years to issue the main tabular data from the last decennial population census in magnetic tape format, with only certain summaries available in printed form. The result has been a reassessment by many government agencies, universities, and private industrial firms of their ability and need to take on census data in this form. Tapes are obviously more complicated to use than books, but they offer certain compensations. For example, if a searcher wanted to know how many persons living in a mile-square area in each of the major cities of the country lived in homes that had television and radio sets but that lacked proper plumbing facilities—and also perhaps what the persons' employment status was at the time of the census—the study might take months to do manually, but it would not be difficult at all for a computer. (See Chapter 28 for further discussion of the use of automated data banks.)

The category of **statistics** traditionally makes one think of economic and sociological studies—and many an engineer will find a need for just such information. There are two outstanding government-prepared handbooks in this field: **Statistical Abstracts of the U.S.** and **Statistical Yearbook**, the latter a United Nations publication. Their coverage is remarkably broad. At the other end of the spectrum, more restricted information, such as price information on certain industries or trades, can frequently be found in periodical articles, and often easily found through periodical indexing services, such as **Applied Science and Technology Index** (see Chapter 13) and **Business Periodicals Index** (see Chapter 25).

Much detailed statistical information is available from the government, often in the form of monthly mimeographed lists, sometimes not for sale but available upon request. The February issue of the **Monthly Catalog of U.S. Government Publications** cites these lists, along with all serial, or periodical-type publications issued by the government and including all the regularly issued statistical compilations available from various agencies.

The Department of Commerce is probably one of the most important government agencies issuing statistical reports of importance to the entire business, agricultural, and industrial sector. The Bureau of the Census alone issues innumerable reports, not only on population counts but on industrial censuses, such as the **Census of Manufacturers**, as well as special reports for each of 50 mineral industries. It also issues the **Census of Business** (mostly nonmanufacturing firms), the **Census of Transportation**, the **Census of Construction Industries**, the **Census of Agriculture**, and the **Census of Governments**. In addition, it issues many reports on very specific topics, such as the **Sale of New One-Family Homes**, a series issued monthly, quarterly, and annually. There is also the Bureau of International Commerce, which is devoted entirely to foreign commerce. A special index to the voluminous output of data from this agency, **Index to Foreign Market Reports**, has listings by country as well as subject.

Certain products are so commonly used by engineers that data about them have been compiled in a special type of literature—**data compilations**. Take, for example, transistorized circuits. On a job involving them, engineers must recommend the exact solid-state components by manufacturer's name and part number, and to get this information they reply on manufacturers' loose-leaf or paperback publications, or on across-the-board compilations sold by commercial publishers. The latter give general information on many manufacturers; products from many sources can be compared in one table or section.

No account of data compilations would be complete without mentioning the **National Standard Reference Data System**, a program carried out since its origin in 1963 under the sponsorship of the National Bureau of Standards (NBS). Its purpose is to provide critically evaluated compilations of data of physical science. To date, compilations have been made, or are well under way, in nuclear properties, atomic and molecular properties, solid-state properties, mechanical properties, and so on. The reports in this series, such as **Tables of Molecular Vibrational Frequencies**, are issued separately. The NBS program has, at least in part, taken over the updating of a monumental seven-volume set which is still well worth knowing about, despite its age—the **International Critical Tables**.

A new format of data compilations and for critical reviews in the physical sciences is the recent **Journal of Physical and Chemical Reference Data**, published jointly by the American Institute of Physics and the American Chemical Society. The journal consists of the output of the **National Standard Reference Data System** but uses the periodical format to reach a wider audience than the report format did and to allow faster publication than government schedules did.

A related function of the NBS is the creation of the **Catalogue of Standard Reference Materials**, a function monitored and coordinated by the Office of Standard Reference Materials. The purpose is to create items for precise standard measurements in the laboratory, such as a standard for the calibration of instruments used in measuring hydrogen in steel. The standards are not necessarily prepared by NBS; its role is to act as sales agent and coordinator of the project.

SELECTED EXAMPLES

Tables

General

CRC Handbook of Tables for Applied Engineering Science. 2d ed. Edited by Ray E. Bolz and George L. Tuve. Cleveland: CRC Press, 1973. 1,150 pp.

> This handbook provides tabular data for various types of engineering (chemical, nuclear, electrical, and mechanical) as well as engineering materials, environmental topics, safety, measurements, and so forth. Both SI and conventional units are used.

Chemistry

Handbook of Chemistry and Physics. Cleveland: CRC Press, 1913- [Annual]

> This compilation of data is an especially valuable reference source. It has thousands of tables giving basic physical and chemical data, plus structural formulas for chemicals, conversion factors, mathematical tables, and so forth.

Lange's Handbook of Chemistry. 11th ed. Compiled by John A. Dean. New York: McGraw-Hill, 1973. Variously paged.

> Almost entirely in tabular form, this work is very comprehensive. The tables give properties of chemical substances grouped into the classes of elements, minerals, organic compounds, inorganic compounds, and industrial materials. There are also chemical analysis tables as well as miscellaneous and numerical ones.

See also **Physics**

Construction

Foster, Norman, ed. **Practical Tables for Building Construction.** New York: McGraw-Hill, 1963. 248 pp.

> This handy pocket-sized book is for the use of architects and

engineers, particularly in the field. Besides general tables (conversion factors, strength of materials, heating, and so on), it has many specialized tables (earthwork, lumber, roofing, piping, and the like).

Mathematics

Burington, Richard S., comp. **Handbook of Mathematical Tables and Formulas.** 5th ed. New York: McGraw-Hill, 1973. 500 pp.

> This volume emphasizes the basic tables (trigonometry, integrals, logarithms, powers, and so forth) as well as interest and actuarial tables, constants, and similar material. It also includes definitions and theorems.

Comrie, L. J., comp. **Barlow's Tables of Squares, Cubes, Square Roots, Cube Roots, and Reciprocals of All Integers up to 12,500.** 4th ed. New York: Barnes & Noble, 1941. 258 pp.

> This revision of a very famous set of tables, dating back over 150 years, is still a very useful compilation of basic data.

Comrie, L. J., comp. **Chamber's Shorter Six-Figure Mathematics Tables.** New York: Wiley, 1972. 389 pp.

> This work concentrates on general tables having four-figure or six-figure values. It includes logarithms, trigonometric functions, exponential and hyperbolic functions.

CRC Handbook of Tables for Mathematics. 4th ed. Edited by Samuel M. Selby. Cleveland: CRC Press, 1975. 1,100 pp.

> The handbook contains hundreds of tables, arranged in groups such as logarithms, trigonometry, matrices and determinants, differential equations, and probability.

CRC Handbook of Tables for Probability and Statistics. 2d ed. Edited by William H. Beyer. Cleveland: CRC Press, 1968. 642 pp.

> This collection of tables involves probability, various distributions (normal, binomial, Poisson, Chi-square, and so on), range, correlation coefficient, quality control, and related topics.

CRC Standard Mathematical Tables. 23d ed. Edited by Samuel M. Selby. Cleveland: CRC Press, 1975. 765 pp.

> The material on measurements has been enlarged in regard to the use of the metric system. Major topics include mensuration,

circular and hyperbolic trigonometry, analytical geometry, algebra of sets, determinants and matrices, integral tables, Laplace and Fourier transforms, vector analysis, etc.

Dwight, Herbert B., ed. **Mathematical Tables of Elementary and Some Higher Mathematical Functions.** 3d ed. New York: Dover, 1961. 337 pp.

This inexpensive paperback set of tables includes not only common tables such as squares and roots but also items such as Bessel functions, hyperbolic functions, and gamma functions.

Fletcher, A., and others, eds. **An Index of Mathematical Tables.** 2d ed. Reading, Mass.: Addison-Wesley, 1962. 2 vols.

An invaluable source for locating mathematical tables, particularly obscure ones not commonly found in handbooks. The index is very comprehensive.

Jahnke, Eugene, and others. **Tables of Higher Functions.** New York: McGraw-Hill, 1960. 318 pp.

Although now out of print, this still remains a very useful book. It is divided into twelve sections, such as error functions, elliptic integrals, Mathieu functions, orthogonal polynomials, etc. Text and headings are in both English and German.

Klerer, Melvin, and Fred A. Grossman. **A New Table of Indefinite Integrals, Computer Processed.** New York: Dover, 1971. 198 pp.

This useful compilation is arranged by similar types.

Peters, Jean. **Eight-Place Tables of Trigonometric Functions for Every Second of Arc.** New York: Chelsea Publishing Company, 1963. 954 pp.

Besides the eight-place tables for sine, cosine, tangent, and cotangent, this volume includes 21-place tables for sine and cosine.

Spiegel, Murray R. **Mathematical Handbook of Formulas and Tables.** New York: McGraw-Hill, 1968. 271 pp.

This paperback compilation includes some 2,400 formulas and 60 tables. Part 1 consists of formulas; Part 2 is made up of tables. Topics range from elementary to advanced.

U.S. National Bureau of Standards. **Handbook of Mathematical**

Functions, with Formulas, Graphs, and Mathematical Tables. Washington, D.C.: Superintendent of Documents, 1964. 1,046 pp. (Applied Mathematics Series No. 55.)

> Aimed at serving scientists and engineers in many fields, the 29 chapters consist mostly of tabular data or formulas ranging from Legendre functions to orthogonal polynomials. It is an outstanding compilation.

Mechanical elements

American Society of Mechanical Engineers. **ASME Handbook: Engineering Tables.** Edited by Jesse Huckert. New York: McGraw-Hill, 1956. Variously paged.

> Essentially a collection of tables often wanted by engineers but not commonly found in handbooks, this work is centered around 15 topics, such as gears, bearings, springs, and piping.

Metals

Smithells, Colin J., ed. **Metals Reference Book.** 5th ed. Reading, Mass.: Butterworth, 1975. 1,150 pp.

> This very comprehensive book contains tabular data on all aspects of metals such as crystallography, diffusion, physical properties, electrical and mechanical properties, welding, and casting.

Physics

Kaye, George W., and T. H. Laby, eds. **Tables of Physical and Chemical Constants and Some Mathematical Functions.** 13th ed. New York: Wiley, 1966. 249 pp.

> The major topics center around general physics, chemistry, atomic physics, and mathematics. This is essentially a collection of tables, with a brief amount of text and formulas. The mathematical section is small.

Purdue University. Thermophysical Properties Research Center. **Thermal Conductivity-Metallic Elements and Alloys.** Edited by Y. S. Touloukian. New York: IFI/Plenum, 1970. 1,469 pp.

> This is volume 1 in a series of 13 volumes dealing with thermophysical properties of materials. It presents extensive data for elements and alloys and intermetallic compounds, and is a very thorough compilation. There is an index by material name.

Smithsonian Physical Tables. 9th rev. ed. Compiled by William W. Forsythe. Washington, D.C.: Smithsonian Institution, 1964. 827 pp. (Publication #4169)

Nine hundred tables present data on a wide scope of physical properties. They include, for instance, atomic data, optical properties, many tables on heat, properties of gases and fluids, and atmospheric data.

See also **Chemistry**

Steam

American Society of Mechanical Engineers. **1967 ASME Steam Tables.** Edited by C. A. Meyer, and others. New York: The Society, 1967. 328 pp.

This book consists of tables of the thermodynamic and transport properties of steam and includes a Mollier chart (enthalpy-entrophy diagram).

Kennan, Joseph H., and others, eds. **Steam Tables: Thermodynamic Properties of Water including Vapor, Liquid, and Solid Phases.** (English units.) New York: Wiley, 1969. 162 pp.

Following a large section of tables, an appendix explains the techniques and problems of compiling the tables. An edition is also available featuring metric units.

Statistics

Chemical and Engineering News. Facts and Figures. Washington, D.C.: American Chemical Society, 1956-   [Annual]

This special supplemental issue presents production figures, price indexes, exports and imports, and related statistics for scores of chemicals. Data cover the three previous years with an estimate for the current year.

Edison Electric Institute. **Statistical Year Book of the Electric Utility Industry.** New York: The Institute, 1928-   [Annual]

The publication covers sales, type of equipment used, and so forth. It consolidates information from many sources, much of it in the form of annual statistical reports sent to the Institute plus government sources.

Metal Statistics. New York: Fairchild, 1908-   [Annual]

TABLES, STATISTICS, AND DATA COMPILATIONS

This publication gives such information as consumption, supply outlook, and recent price trends for all types of metals. It has extensive coverage for such major industries as aluminum, copper, pig iron, and steel and includes listings of major trade and industry associations, as well as a directory of "Where to buy and Where to Sell."

Statistical Yearbook. New York: United Nations, 1948-   [Annual]

This compilation of statistical tables treats a wide range of topics on an international basis. It has an index to the contents arranged by country.

U.S. Department of Commerce. Washington, D.C.

The department is probably the most prolific source of publications concerning economic, financial, and technical statistics. It is both domestic and international in scope. Various pamphlets and lists can be cited to indicate the great number of publications available. Among them are the following:

Business Service Checklist. This weekly listing of recent departmental publications is available on subscription at a very nominal price.
Selected Publications to Aid Domestic Business and Industry. This description of many periodicals issued by the department ranges from data on the copper industry to population characteristics.
Checklist of International Business Publications. The checklist features a listing by country of the publications available for each region.

U.S. Department of Commerce. **Statistical Abstract of the United States.** Washington, D.C.: Superintendent of Documents. 1878- [Annual]

One of the best and least expensive sources of statistics about the United States, this abstract covers a wide range of topics. It is essentially a collection of many tables, with a subject index.

Wasserman, Paul, and others, eds. **Statistics Sources.** 3d ed. Detroit: Gale Research Company, 1970. 387 pp.

This alphabetical list of thousands of subjects, such as manganese-import prices or the Belgian employment and labor force is designed to help the user find current statistical data. For each topic, at least one current source of statistics is listed.

Data Compilations

ISA Handbook of Control Valves. Edited by J. W. Hutchison, Pittsburgh: Instrument Society of America, 1971. 308 pp.

The handbook includes not only the principles of valve design but also detailed information about available products, including manufacturers' model numbers.

Journal of Physical and Chemical Reference Data. New York: American Institute of Physics, 1972-   [Monthly]

Jointly published by AIP and the American Chemical Society, the purpose of the journal is to serve as a quick means of publishing information from the **National Standard Reference Data System.**

Lenk, John D. **Practical Semiconductor Databook for Electronic Engineers and Technicians.** Englewood Cliffs, N.J.: Prentice-Hall, 1970. 260 pp.

The book presents equations and descriptions on the use of semiconductors, including field effect transistors, integrated circuits, semiconductor power supplies, and other related topics.

National Research Council. **International Critical Tables of Numerical Data: Physics, Chemistry, and Technology.** New York: McGraw-Hill, 1926. 7 vols.

Although this set is out of print, it is still a useful set of critically evaluated data in the physical sciences and engineering. It has a detailed subject index. The sources of data are carefully cited.

National Standard Reference Data System. Washington, D.C.: Superintendent of Documents, 1963-

The U.S. National Bureau of Standards sponsors this project, whose object is to promote the development of data relating to physical science. The data are to be critically evaluated and carefully checked for accuracy. The system was established in 1963 and then strengthened in 1968 by the passage of the National Standard Reference Data Act. The resulting publications are on sale at the Superintendent of Documents, Washington, D.C.

U.S. National Bureau of Standards. Office of Standard Reference Materials. **Catalog of Standard Reference Materials.** Washington, D.C.: Superintendent of Documents, 1900-   [Biennial] (NBS Special Publication no. 260.)

This index to over 600 standard reference materials, such as nonferrous alloys, cast iron, cements, and carbides, gives chemical compositions as well as physical descriptions. Materials for calibrating test equipment are also included. It facilitates measurements on materials and also the controlling of industrial production processes. Prices are given in a semiannual supplement.

CHAPTER 20

Patents and Trademarks

A **patent** protects an inventor by granting him or her exclusive rights to a particular design or product for a specific length of time and in a specific country. In the United States, most patents are issued for seventeen years. The Patent Office, which granted only three patents during its first year of operation (1790), has, in recent years, grown to the point where the annual total is around 50,000. By 1975 the grand total of patents issued was around 4 million.

This country requires that the inventors themselves apply for the patent, although they may turn over ownership to their employing company—a common custom. (To balance this arrangement, many companies give patentees bonuses, public recognition, or promotions—and have, of course, given the money, equipment, and other support necessary for the work in the first place. The day of the lone basement inventor has not quite disappeared, but it is no longer the usual pattern—it is too expensive.) The patent indexes now issued by the U.S. government leave much to be desired, however, and the services of a patent attorney are usually needed if one wants to determine whether or not a certain type of item has already been patented. The indexes are not thorough, and the language is intentionally vague—exact details would give away too much special knowledge. Patent reading is definitely not for the neophyte, but is for the knowledgeable person with a strong background in the field. Nevertheless, perhaps a dozen or so public libraries across the country have complete sets of patents.

Many of the standard periodical indexing services, such as **Chemical Abstracts**, include patents; this particular abstracting service publishes a special index arranged by patent numbers, for both United States and foreign patents. CA indexing is also competent and thorough. In addition, patent indexing is done by commercial firms that usually specialize in particular subjects, such as plastics or pharmaceuticals. The quality of this work varies considerably. Some such services also cover foreign patents.

A product or service is often best known to the public by an identifying design or slogan. This design or slogan, too, can be protected—by a **trademark**, issued by the Patent Office, which gives the owner exclusive rights to its use for twenty years, and is then renewable at the end of each following twenty-year term. As with patents, the protection applies only to this country. (Foreign registration is required for protection outside the United States.) The first trademarks—121 of them—were registered in the United States in 1870; around 20,000 a year are now registered.

Except for the index published by the Patent Office, little or no indexing of trademarks is done in this country.

SELECTED EXAMPLES

U.S. Patent Office. **Directory of Registered Patent Attorneys and Agents Arranged by States and Countries.** Washington, D.C.: Superintendent of Documents, 1900-

> The listing of patent attorneys and agents registered to practice before the U.S. Patent Office is arranged geographically.

U.S. Patent Office. **Manual of Classification.** Washington, D.C.: Superintendent of Documents, 1900-

> This loose-leaf volume, available on a subscription basis for supplements, lists the numbers and titles of the more than 300 main classes and the 66,000 subclasses used in the subject classification of patents.

U.S. Patent Office. **Official Gazette: Patents.** Washington, D.C.: Superintendent of Documents, 1971-   [Weekly]

> The listing of recently issued patents shows a selected diagram, a brief abstract of the text, the name of the patentee, and so forth. It has an annual index of patentees and of subjects. Issued since 1971, it continues earlier versions that date back to 1872.

PATENTS AND TRADEMARKS

U.S. Patent Office. **Official Gazette: Trademarks.** Washington, D.C.: Superintendent of Documents. 1971-   [Weekly]

This index to recently issued trademarks gives an illustration of the trademark. There is an annual index of registrants. Issued since 1971, it continues earlier versions that date back to 1872.

CHAPTER 21

Dissertations and Masters Essays

Doctoral dissertations constitute an unusual type of information: because of the requirement for uniqueness before a topic is accepted, they are often at the advance fringe of research at the time of writing. On the other hand, by the time many of them become known outside the authors' circles, the writers may have submitted the main ideas to journals for quicker, more widespread publication.

Even when one knows about a given dissertation, a few are hard to obtain. For example, some of our leading universities have not, for reasons of their own, seen fit to participate in the microfilming of dissertations that is done nationally by a private company—University Microfilms—which then makes them available for sale to the public. Those dissertations that **are** microfilmed, however, are then usually not otherwise available from the university unless they were written before the school joined the program, in which case copies can be sold by the university. In either event, once a dissertation of interest is obtained, it can be most detailed and worthwhile and is usually well documented.

The master's thesis has much less standing than the doctoral dissertation, which is probably a just evaluation when one realizes the differences in time, originality, and amount of research expected for the two types. In many universities the requirements for a master's essay, as they are often called, are gradually being abandoned, so their number has declined. Indexes for them exist, but records concerning them lack cohesion compared with those for dissertations. Again, the libraries at the universities where they were written often have the only copy, and local custom dictates whether or not additional copies can be made or interlibrary loans granted.

SELECTED EXAMPLES

American Doctoral Dissertations. Ann Arbor: University Microfilms, 1955/1956-   [Annual]

This is a complete listing of all doctoral dissertations accepted by United States and Canadian universities, and thus includes a number that are not listed in **Dissertation Abstracts International.** Listings are arranged by school, under appropriate subjects. There is also an author index.

Dissertation Abstracts International. Part B: The Sciences and Engineering. Ann Arbor: University Microfilms, 1938-   [Monthly]

This index to doctoral dissertations accepted by over 350 United States and Canadian universities is soon to include European institutions. It is arranged by broad categories, such as engineering and chemistry. Very detailed abstracts are given. Each issue has a key word, title, and author index; the author index cumulates annually. Copies of dissertations are offered for sale.

Comprehensive Dissertation Index, 1861-1972. Ann Arbor: University Microfilms, 1973. 37 vols.

There is bibliographic information on over 400,000 dissertations, arranged by broad subjects, then alphabetically by key words. There are author and key-word indexes. Separate volumes are available for specific disciplines.

Masters Theses in the Pure and Applied Sciences Accepted by Colleges and Universities in the United States and Canada. New York: Plenum, 1955-   [Annual]

This list provides information about the masters' theses from over 200 United States universities and colleges. It is arranged by subject categories, such as mechanical engineering and astrophysics. Formerly published by Purdue University, Thermophysical Properties Center.

CHAPTER 22

Newspapers

One problem with newspapers—general, specialized, daily, or weekly—is that of finding information in back (older) issues, and sometimes in a hurry. One of the few newspapers to be completely indexed is **The New York Times,** and its coverage of national and international news of all types gives the index considerable scope. An indexed reference to a national event indicates when that event would have been cited in other newspapers, thus giving a lead to items in other papers—including the specialized ones.

Unfortunately, most specialized papers are not indexed. Even an item of great importance must be located by first finding it in general newspapers or trade magazines (by using the indexing services that index these publications) and then, by taking the date of publication, finding the issue of the specialty paper most likely to carry the item.

Engineers interested in management and finance might well want to become regular readers of a newspaper such as **The Wall Street Journal.** It is, of course, a hybrid. It includes a good coverage of general news as well as feature articles and tabulations that pertain strictly to business and finance.

SELECTED EXAMPLES

American Metal Market. Somerset, N.J., 1882-   [Daily (weekdays)]

> Published every business day, this newspaper includes news affecting production and sales of metals and metal products, as well as prices and trend indicators.

Chemical Marketing Reporter. New York: Schnell Publishing Company, 1871- [Weekly]

This newspaperlike publication, formerly known as **Oil, Paint and Drug Reporter**, is a source of detailed information about the prices of chemicals, production statistics, personnel news, new products, and so on.

New York Times Index. New York, 1851- [Semimonthly]

The index, which has annual cumulations, has entries under people, places, and subjects. The publication is very carefully indexed, with many cross-references, and is found in many libraries. For the past two decades it has been available on microfilm. Also, within the last five to ten years a magnetic tape version has been available, supplemented by data from other sources such as other newspapers and periodicals.

The Wall Street Journal. New York, 1889- [Daily (weekdays)]

The newspaper specializes in business and financial news, but also has local, regional, national, and international coverage of general news.

CHAPTER 23

Translations

The English-speaking world has long had a poor reputation as far as learning other languages is concerned, yet it is essential for American engineers to keep up with the work of scientists in other countries. What to do? Using a professional translator is expensive.

A few alternatives are available. One is to use some of the dozens of important periodicals that are regularly translated cover to cover. These are published by commercial firms or under thy sponsorship of professional engineering or technical societies. Russian and Japanese journals are those most commonly translated completely. Also, as described in **Translation Register-Index**, thousands of translations are available from a national pool in Chicago, whose translations are well indexed.

Another alternative is to use the English-language summaries that are sometimes found inside a periodical. Or, in an emergency, one can study whatever nonverbal material there is and also reach for the appropriate dictionary. (Dictionaries are discussed in Chapter 4.)

The translation of technical material is simplified somewhat by the fact that most of the foreign articles of greatest value appear in a few key languages—Russian, German, Japanese, and French, in that order, made the most contributions to chemical literature in 1970.[1]

---

[1] Dale B. Baker, "World's Chemical Literature Continues to Expand." **Chemical and Engineering News** 29(28):37-40, 12 July 1971.

SELECTED EXAMPLES

Himmelsbach, Carl J., and Grace E. Brociner. **A Guide to Scientific and Technical Journals in Translation.** 2d ed. New York: Special Libraries Association, 1972. 49 pp.

This work presents a list of journals which are being translated into English on a cover-to-cover basis. It includes both translated and original titles.

Translation Register-Index. Chicago: National Translation Center, John Crerar Library, 1967-   [Monthly; semiannual cumulations]

The TRI indexes translations available at a nominal fee from the center. Each issue has patent and journal citation indexes, which cumulate quarterly and annually. It has been published since 1967 under its current title, but continues previous indexes having various titles. At one time the project was operated by the Special Libraries Association.

World Index of Scientific Translations and List of Translations Notified to ETC. Delft, Netherlands: European Translation Centre, 1967-   [Monthly]

This listing of available translations covers periodical articles, patents, and standards. It is arranged by the COSATI subject classification system. A supplement lists journals translated cover to cover, citing over 1,100 titles.

CHAPTER 24

Maps, Atlases, and Engineering Drawings

Maps have many uses, the most obvious one being to serve their various specific purposes. A geological map would not serve a marketing manager at all, nor would a road map be of much use to a civil engineer needing data about rock formations.

Other important points are the scale, which must be appropriate (trying to find a small town on a map of the entire United States is an obvious example of this), and, frequently, the publication date. Some types of maps—political, for example—can become outdated fairly quickly.

Another feature to consider is type of projection used—the manner in which the map is constructed, which is an involved subject in itself. Part of the problem of projection is the basic difficulty of trying to show a curved surface (the Earth) on a flat surface (the map). Most users are so familiar with the usual Mercator projection that any other, such as a polar projection, seems distorted. Yet a navigator in a polar region would be in trouble if only a Mercator projection were at hand, because it drastically distorts those regions.

The best maps are usually published in numbered series by important government or scientific agencies. The government sources may be federal (the largest source in this country), state, or local government. Two important sources of federal maps are the U.S. Geological Survey and the Defense Mapping Agency. Technical and scientific societies are responsible for many maps in particular subjects. Commercial sources usually concentrate on road maps and political maps.

The best sources of information about the availability of particular maps are the lists issued by the map publishers themselves. Difficulties may result from not knowing which agencies (or publishers) to turn to, or, once the lists are located, finding a map library or collection that has the maps, particularly if you need them quickly.

Universities can often help, as can map curators for major societies and agencies issuing maps. And one good source does exist to help find good map collections—**Map Collections in the United States and Canada.**

Atlases—bound sets of maps—also range from the general to the highly specific. Even the moon has atlases devoted to it.

Closely related to maps are charts, such as those showing navigational hazards as indicated by depth soundings. Again, most of them are government-produced and -distributed. A close cousin to a map or chart is an aerial photograph, usually taken from aircraft, with the exception now of those sent back by satellites, which have been valuable in providing photographs of meteorological conditions. Military intelligence is also becoming increasingly dependent on infrared and electronic data provided on a global scale by special satellites. Certain civil engineering projects, such as highway design and city planning, are also vastly simplified by the use of aerial photographs.

Great quantities of drawings are involved in many engineering projects. In most cases these drawings are available only in limited quantities and then only from those responsible for the project. In competitive projects, access to drawings may be especially carefully guarded—by the government and/or company involved—to avoid unauthorized disclosure of information.

In some cases the problems of handling vast quantities of drawings have led to microfilming the originals to expedite handling and reduce costs. A system now in common use relies on what are called aperture cards. These cards are the same size as ordinary accounting-machine punched cards but have an opening, usually on the right side, large enough for a frame—generally 35 mm—of microfilm. The cards can be keypunched for sorting or searching to find a drawing with a particular part number, contract number, or title. Full-size copies can be made by using simple pushbutton-operation microfilm printers.

One article describing the uses of microfilm for engineering drawings mentioned an actual case in which this system saved over $3,200 for a single project.[1] This same article touched on what is still a relatively new practice—using computers to prepare engineering drawings. So far, these drawings are economically feasible only if they can be used for several years, but no doubt changes will be made as this system gets further study, as is indicated by a more recent review of all the engineering uses of microforms, including applications to drawings and reports.[2]

SELECTED EXAMPLES

Cosmopolitan World Atlas. Enlarged "Planet Earth" ed. Chicago: Rand McNally, 1971. 428 pp.

> This general reference atlas has more than 200 pages in full color. The index has 82,000 entries. Although it is moderately priced, less expansive atlases are also available.

International Maps and Atlases in Print. New York: Bowker, 1974. 862 pp.

> The work describes over 8,000 maps and atlases, giving full details as to title, scale, series number, editor, size, and source of publication. It is arranged geographically and ranges from star charts to railway maps.

National Atlas of the United States of America. Washington, D.C.: U.S. Geological Survey, 1970. 417 pp.

> The atlas includes topography, geology, climate, and water resources in the physical features section. It also encompasses historical and political features. There is an index of over 41,000 place names, and 336 pages are multicolored maps. This valuable reference tool required eight years of planning to produce.

Special Libraries Association. Geography and Map Division. Directory Revision Committee. **Map Collections in the United States and Canada.** Edited by David K. Carrington. New York: 1970. 159 pp.

[1] W. P. Southard, "Designing with Microfilm." **Chemical Engineering** 78(11):133-138, 17 May 1971.
[2] Don Mennie, "Engineering Micrographs." **IEEE Spectrum** 12(10):62-66, Oct. 1975.

This work describes the size and type of map collections in over 600 institutions, of which 43 are Canadian. The institutions are arranged alphabetically by city within a state or province. Subject specialization is described.

CHAPTER 25

Business Information

Many engineers working in private industry will eventually need to know about the management end of business—a subject which may be new to them and which can be extremely time-consuming to learn and to keep up with. The federal government is the source of much business information, such as procurement regulations for various agencies, or explanation of tax laws, but most companies also use commercial publications for more detailed explanations of these laws and regulations. Three specific types of services are available.

One type of service, for managers, is a digest and analysis of the highlights of pertinent legislation and administrative rulings. There are frequent updating supplements. Some of these services are tailored to particular industries.

A second type of service, for those in the financial and investment aspects of business, deals with records of credit standing or values of companies' stocks and bonds. Some of these are also geared to particular types of businesses.

A third type of service, for executives, is the newsletter, which gives highlights of new ways to improve a company's financial record. Closely related to this are the newspapers devoted chiefly to business and finance, such as **The Wall Street Journal.**

A related reference tool is the index to business periodicals. Although there are several in this field, **Business Periodicals Index** is particularly popular in business libraries.

SELECTED EXAMPLES

Publications

Business Periodicals Index. New York: Wilson, 1958- [Monthly]

The index covers the fields of accounting, advertising, public relations, automation, banking, communications, economics, finance, investments, labor, management, marketing taxation, and specific industries, trades, and businesses. It is restricted to English language journals; over 150 are indexed completely. There are no abstracts or other indexes. It includes much in the way of news items often hard to locate otherwise such as biographical notices and price statistics. Very useful in spite of no annotations. It cumulates quarterly and annually.

Chemical Industry Notes. Columbus, Ohio. American Chemical Society, 1972- [Weekly]

This publication gives concise, factual news items of interest to those in the business end as well as the technical sector of chemically based industries. It scans both periodicals and newspapers on an international basis. Each issue has a key-word subject index. Topics include production trends, pricing, marketing, manpower, labor activities, and so on.

Publishers of Services

Commerce Clearing House, Inc. Chicago.

This service presents reports on tax and business law in over 100 special areas, in loose-leaf format with supplements. A wide selection is possible. For instance, services range from SEC rules on accounting practices to explanations of how to handle government contracts. One is devoted to pollution control, another to aviation laws and regulations. Guidebooks that summarize laws in over 40 areas are also available.

Moody's Investors Service, Inc. New York.

This service concentrates on investment information. It issues six loose-leaf services concerned with data on industrial firms, public utilities, transportation, banks and finance, municipal agencies, and government agencies. Thousands of organizations are covered, chiefly in terms of financial status (income, debts, stock and bond descriptions, and the like). Weekly newsletters are also available, covering the stock and bond markets. Other manuals give the background and performance of common stocks and bonds.

BUSINESS INFORMATION

Prentice-Hall, Inc. Englewood Cliffs, N.J.

The service offers over three dozen specialized business and legal services, in loose-leaf form, which range from social security law to property taxes, from wills to natural resource taxes, from banking to pensions. Supplements keep the books up to date. The style varies from verbatim text of court decisions to those written in nontechnical, easy-to-understand language. One series is written for executives, to aid in cost cutting and in avoiding errors of management.

Standard and Poor's New York.

Provides a variety of publications from loose-leaf services to charts to punched cards and magnetic tapes. Emphasis is on investment information. The tapes, for example, give prices for listed and unlisted securities, with the choice of format ranging from daily to quarterly summaries. Dividend records, published daily, weekly, or quarterly, are offered, as well as loose-leaf sets of reports on all companies listed on the N.Y. Stock Exchange. **Poor's Register of Corporations, Directors and Executives** lists over 260,000 officials in some 33,000 companies.

CHAPTER 26

Quotations Indexes and Thesauri

Many people use quotations to make a point or to sound a particular tone in a speech or article. Probably most of them used a quotation reference book to check their memories or to find an appropriate quotation in the first place. (Certainly most people in all fields have heard of **Bartlett's Familiar Quotations**.) These books contain an index to outstanding citations from the world's literature and are simple to use. The index is usually arranged by key word, or, in the case of a concordance, by even less-important words.

Thesauri come in two different types. One has been a standard reference tool for many decades; the other is more or less a child of the post-World War II days and originated as the by-product of the growth of technical report literature.

The popular thesaurus is essentially a guide to synonyms or antonyms. Some thesauri are arranged with the key words in alphabetical order, followed by the synonyms and, for more important words, the antonyms. Related terms are cross-referenced. Another style of organization is by broad categories, under which an alphabetical listing appears. This second style occasionally leads to confusion because a word can sometimes appear in several places.

Technical thesauri are completely different. They are concerned with scientific and engineering terms and their main use is to indicate which terms to use for a search for information from an index, a computer system, or a reel of magnetic tape. For information on lasers, for example, looking under **optical masers** would give one a cross-reference to **laser**; in addition, under **laser** there might be related terms, such as **quantum electronics**, various types of lasers in the system, and perhaps names of devices using lasers. Since virtually limitless related terms could be built into a technical thesaurus, some sort of arbitrary limit is generally set at the beginning.

Thesauri of this type must of course be kept up to date.

SELECTED EXAMPLES

Quotations

Bartlett's Familiar Quotations. 14th ed. Edited by E. M. Beck. Boston: Little, Brown. 1968. 1,750 pp.

This book is the best known source of quotations identification.

Thesauri

Engineers Joint Council. **Thesaurus of Engineering And Scientific Terms.** New York: The Council, 1967. 690 pp.

The thesaurus is best described by the subtitle: "A list of engineering and related scientific terms and their relationships for use as a vocabulary reference in indexing and retrieving technical information." Besides the main alphabetical approach, there is also a key-word index, an index by the 22 COSATI fields, and an index showing the descriptors arranged by family relationships.

The Original Roget's Thesaurus of English Words and Phrases. New ed., rev. and modernized by Robert A. Dutch, New York: St. Martins, 1965. 1,405 pp.

This is the standard work for identifying closely related words or opposite terms in common, nontechnical use.

CHAPTER 27

Personal Contacts, Including Professional, Industrial, Educational and Governmental Sources of Information

Studies have shown that much of what engineers and scientists learn comes from communicating with friends and colleagues—sometimes called the "invisible college," or, by the British, the "old boy" circle. Some of this communication takes place at technical symposia and conferences where, often, more is learned informally than from the formal meetings and the papers being delivered.

Another major way of locating information is through professional societies. Some of the directories mentioned in Chapter 10 list such organizations (for example, **The Encyclopedia of Associations**).

There are literally thousands of such groups, some with only a few hundred members and perhaps highly specialized, others with tens of thousands of members and in a general subject area, such as electrical engineering or chemistry. Their headquarters invariably have knowledgeable staff members who can either answer questions or offer suggestions for printed materials or personal contacts.

A major concentration of organizations is in the United Engineering Center in New York City. It houses the headquarters of several leading engineering societies in the United States, and is in the same building as one of the best engineering libraries in the country, the Engineering Societies Library, which they sponsor. This library will loan materials to members of the sponsoring societies and has a reading room open to serious users.

Reference librarians in general are also available for personal assistance. Many of them have strong backgrounds in engineering. A few large technical libraries even, for a fee, offer a search service. The personal contact is limited, but it is the link between the engineer and the literature.

A large part of a country's technical skills and knowledge is found within the industrial sector, where research and practical application are often developed into amazing degrees of specialized knowledge. The only trouble is that much of this information is company- or government-classified and therefore impossible to get. On the other hand, much of it is common enough knowledge so that employees are allowed to talk about it to inquirers. A telephone call or letter can tell you. Or, if applicable company reports might exist, a call or letter to the library or information center might be in order.

Another great concentration of technical knowledge is found in colleges and universities. Theoretical research is often emphasized, and industrial firms and government agencies frequently hire faculty members of these schools as consultants.

Many university projects are sponsored by government agencies. This means that reports must be submitted to the agencies, and many end up in the major report indexing services and their respective sales agencies.

Some federal government agencies are also set up to answer questions from individuals. One of these, the National Referral Center, is part of the Library of Congress. It cannot do exhaustive searches, but it can usually provide the name of someone who will be willing to give some quick answers.

In recent years some states have maintained agencies to help the private inquirer use the huge amount of government information, but the current trend has been to dissolve many of these worthy agencies because of economic stringencies.

The discussion of machine-readable data in Chapter 28 includes a description of several services offered by government agencies, and supplements this section. Chapter 19 also includes information about special services available from government agencies.

SELECTED EXAMPLES

U.S. Library of Congress. National Referral Center. Washington, D.C.

This division serves the general public by directing inquirers to the best sources for answers to their questions. Both technical and nontechnical inquiries are handled. The center may refer inquiries to individuals or to organizations. There is no charge for this service. Inquirers may call, write, or ask their questions in person, and they will be given the name, address, phone number, and a brief description of the source recommended by the center. Directories are issued (see next item).

U.S. Library of Congress. National Referral Center. Science and Technology Division. **A Directory of Information Resources in the United States: Physical Sciences, Engineering.** Washington, D.C., 1971. 803 pp.

The directory lists nearly 2,900 sources of information on the physical sciences and engineering. Information resources listed include private industry, governmental agencies, and educational institutions. Areas of interest are given, as well as size of collection and services available. There is a subject index. This edition supersedes the 1965 edition.

CHAPTER 28

Special Services and Materials (Current Awareness Services; Retrospective Searching; Electronic Data Processing)

In recent years several new types of information service have come into being, chiefly as a result of the introduction of modern data processing equipment to the transferring or retrieval of information. One of the new services is that of keeping engineers supplied with current material to match their needs—**current awareness services**. Another major concept is that of using computers or other data processing equipment for the retrieval of specific information, upon request—**retrospective searching**.

There are three types of current awareness services: a traditional service, using printed publications; a newer service, using machine-readable data banks; and a third service, using audiovisual material.

One type of popular printed publication collects the information from the contents pages of forthcoming periodicals and prints the collected page reproductions in periodical form, usually monthly. An author index may accompany each issue, and occasionally a cumulative index shows in which issue a particular contents page can be found.

The last few years have seen the creation of large machine-readable technical information files stored on magnetic tapes or discs. For many years most of these have been used to provide current awareness service for engineers and scientists. The participating engineers fill out questionnaires which create a profile of their needs. The computer then matches current incoming material to these needs. Those items which match the user's profile are automatically printed out, often in card form, usually on a weekly basis. The output consists of the bibliographic data needed to identify the literature and often includes a brief abstract. Most of these services concern periodical articles and patents, although a few also include technical reports.

Another name for such current awareness systems is Selective Dissemination of Information, or SDI. Much of the success of SDI systems depends upon the care with which the profile is drawn up.

Audiovisual aids make up the third group of current awareness services. One company has made extensive use of audio cassettes for over a year to keep its top executives informed of recent developments in areas of company interest.[1] This service does not replace the traditional SDI service involving journal literature, but it can be tailored to meet specific needs. The service is sometimes called an audio journal.

Several societies, including the American Association for the Advancement of Science and the Institute of Electrical and Electronic Engineers, have issued such items as papers and proceedings in audio cassette form, as have some periodicals, such as **Fortune** and **Nation's Business**, the latter two chiefly for business summaries. Banks and government agencies are also entering this field. The use of video cassettes as a learning tool is still in its infancy.

Retrospective searching is the locating of information not necessarily current. Most such searches are done only when the engineer initiates them through a library or information center. This type of searching was once done only manually, but in recent years experiments have been done to try to overcome the expense of performing these searches with computers, such as by packing the data more densely in the memory units. A breakthrough of this kind is needed in order to reduce the time involved in manual searches of reference tools.

[1]Hanford, W. E., et al. "The industrial chemist and chemical information; the human ear as a medium." **Journal of Chemical Documentation** 11(2):68-69, May 1971.

The agencies listed (see Selected Examples) as issuing machine-readable files for SDI systems are in general the same ones concerned with the possibility of offering retrospective searching. Some are designed for selected clients; others are not.

At least one government agency, the Department of Commerce's National Technical Information Services, has established a search service for private citizens to use. It involves the use of the 325,000 reports (as of 1975) in its computer files, dating back to 1964. Over 50,000 reports are being added annually. Their printed index is described in Chapter 15 (**Government Reports Announcements and Index**). All areas of business, science, and technology are included in their report holdings.

An agency which has a wealth of unusual information is the Smithsonian Science Information Exchange (SSIE), now operated by the Smithsonian Science Information Exchange, Inc., for the Smithsonian Institution in Washington, D.C. Its data bank consists of the descriptions of over 100,000 government-sponsored research projects. The data are all in machine-readable form and can be searched by their computer. The scope now includes all basic and applied research in the life, physical, social, and engineering sciences. In October 1971 an announcement paper, the **SSIE Science Newsletter**, was initiated, listing prepackaged searches ready for the public. It is also available on magnetic tape.

More and more data bases can be searched on-line, particularly through the auspices of commercial firms offering a variety of data bases from which to select, as well as through not-for-profit service centers, such as those operated by the University of Georgia in Athens, Georgia, or the Illinois Institute of Technology in Chicago.

Electronic data processing and the automation of information services are of course relatively new compared to traditional manual methods, and there will probably be a combination of conventional materials and newer methods for many decades to come.

SELECTED EXAMPLES

Chemical Titles. Columbus, Ohio: American Chemical Society, 1960- [Biweekly]

The journal lists the titles of articles appearing in about 700 of the top journals covered by **Chemical Abstracts** and is arranged by the KWIC (Key Word In Context) format. It has an author index and a bibliography that lists the bibliographic details of the items appearing in that issue of CT. It is also available on magnetic tape on a leased basis. The publication serves primarily as a current awareness tool.

Computer Program Abstracts. Washington, D.C.: Superintendent of Documents, 1969-   [Quarterly]

The publication cites computer programs developed by or for NASA, the Department of Defense, and ERDA. The list is divided into six sections; the first describes the programs under 34 subject categories, and the five others are indexes (by subject, originating source, program number, equipment requirements, and accession number). All programs given are available.

Contents of Contemporary Mathematical Journals and New Publications. Providence, R.I.: American Mathematical Society, 1968-   [Biweekly]

This work is a listing of titles of articles from dozens of selected mathematics journals. The articles are arranged by subject, rather than by journal. There is an author and journal index in each issue. Also included are selected new books as well as journal articles.

Current Contents: Engineering and Technology. Philadelphia: Institute for Scientific Information, 1970-   [Weekly]

This compilation of the contents pages of over 700 periodicals covers all aspects of engineering and technology. Each issue has an index to names and addresses of authors. It is also published for other disciplines, such as physical sciences, life sciences, and chemical sciences. A service for obtaining tear sheet copies of articles is available.

Current Papers in Electrical and Electronics Engineering. London: Institution of Electrical Engineers, 1964-   [Monthly]

Jointly published with the Institute of Electrical and Electronics Engineers, this listing of new periodical articles, books, conference proceedings, technical reports, patents, and so forth is arranged by a subject classification system. It has no annotations or indexes. Emphasis is on a current awareness service.

Schneider, John H., and others. **Survey of Commercially Available Computer-Readable Bibliographic Data Bases.** Washington, D.C.: American Society for Information Science, 1973. 181 pp.

This work lists major features of 81 data bases issued by 40 United States and 15 foreign commercial firms. Most are related to science and engineering. Based on November 1972 data.

Computerized Data Bases

The situation concerning the availability of tape services is changing so rapidly that it seems best simply to list some of the data bases in current use for computerized searching of engineering literature. This list is by no means complete—merely representative.

Name	Sponsor (owner)	Number of entries (as of 1975)	Earliest year
BIOSIS PREVIEWS	Biosciences Information Service	1,500,000	1969
CHEMCON	American Chemical Society	1,860,000	1970
COMPENDEX	Engineering Index	400,000	1970
GEO-REF	American Geological Institute	250,000	1967
INSPEC	Institution of Electrical Engineers (a) Physics (b) Electronics and Computers	900,000	1969
MEDLARS	National Library of Medicine	2,350,000	1964
METADEX	American Society for Metals	250,000	1966
NTIS	National Technical Information Service	325,000	1970
SCISEARCH	Institute for Scientific Information	Over 800,000	1974
SSIE	Smithsonian Science Information Exchange	110,000	1974

Index

Abbreviations Dictionary, 30
Abstracting services, 107
Abstracts, 24, 107
Acronyms and Initialisms Dictionary, 30
Advances in . . . . . series, 98
Advances in Chemistry Series, No. 30, 6
Aerial photographs, 164
Aerospace Industries Association of America, 95
Aerospace Year Book, 95
Agricultural Engineers' Handbook, 50
Air Pollution Abstracts, 118
Airport Noise Pollution: A Bibliography of Its Effects on People and Property, 27
Alford, M. H. T., 38
Alford, V. L., 38
Aluminum, 68
American Book Publishing Record, 24
American Chemical Society, 6
American Doctoral Dissertations, 158
American Gas Association, 60
American Institute of Physics, 70
American Institute of Steel Construction, 54
American Institute of Timber Construction, 76
American Men and Women of Science, 88
American Metal Market, 159
American National Standards Institute, 134
American Society for Engineering Education, Engineering School Libraries Division, 81
American Society for Metals, 66
American Society for Testing and Materials, 93, 135
American Society of Heating, Refrigerating and Air-Conditioning Engineers, 51
American Society of Mechanical Engineers, 66, 147, 148
American Society of Tool and Manufacturing Engineers, 66, 67, 71
American Water Works Association, 77
American Welding Society, 67
Amstutz, G. C., 35
Annual Review of . . . . . series, 99
Annual review series, 97-99
Anthony, L. J., 81
Applied Mathematics Series No. 55, 147
Applied Mechanics Reviews, 116
Applied Science and Technology Index, 111, 142
Argall, G. O., 41
ASCE Publications Abstracts, 113
ASHRAE Handbook and Product Directory, 51
ASHRAE Handbook of Fundamentals, 51
ASME Handbook: Engineering Tables, 147
ASME Handbook: Metals Engineering—Design, 66
ASME Handbook: Metals Engineering—Processes, 66
ASME Handbook: Metals Properties, 66
ASTM Standards, 134
ASTME Die Design Handbook, 66
Astronautical Multilingual Dictionary, 40
Atlases, 164
Atomindex, 122, 123
Audiovisual aids, 178
Aviation and Space Dictionary, 31

Baker, Robert F., 60
Ballentyne, D. W. G., 30
Barlow's Tables of Squares, Cubes, Square Roots, Cube Roots, and Reciprocals of All Integers up to 12,500, 145
Bartlett's Familiar Quotations, 171, 172
Barton, David K., 74
Baumeister, T., 65
Beck, E. M., 172
Bell, Harold S., 69
Bennett, H., 31
Berger, Carl, 56
Besancon, Robert M., 47
Best's Safety Directory: Safety, Industrial Hygiene, Security, 138
Beyer, William H., 145
Bibliographic Guide for Editors and Authors, 104
Bibliographies, 23-27
Bibliography of the History of Electronics, 86
Bibliography of the History of Technology, 86
Bilingual dictionaries, 36-38
Biographical information, 87-89
Biological Abstracts, 112
BioResearch Index, 112
BIOSIS Previews (data base), 181
Birnbaum, Max, 93
Black Engineers in the United States, 88
Bland, William F., 70
Bolz, Ray E., 144
Bolz, Roger W., 66
Bond, Richard G., 59
Bones, R. A., 33
Books in Print, 24, 25
Bottle, R. T., 81, 82
Bourton, Kay, 26
Boyd, John, 62
Brady, George Stuart, 46
Brief Guide to Sources of Metals Information, 84
British Union—Catalogue of Periodicals, 105, 106
Broadbent, D. T., 38
Brociner, Grace E., 162
Bronshtein, Ilia N., 64
Brown, Peter, 105
Building Construction Handbook, 55

Building Science Abstracts, 113
Bunshah, R. F., 67
Burger, Erich, 38
Burgess, R. A., 54
Burington, Richard S., 145
Burkett, Jack, 83
Business information, 167-169
Business Periodicals Index, 142, 168
Business Service Checklist, 149

Cagle Charles V., 71
Cagnacci-Schwicker, Angelo, 38
Callaham, Ludmilla I., 38
Card catalog, 16-19
Carrington, David K., 165
Carroll, J. M., 56
Carson, Gordon B., 72
Carter, Ciel M., 82
Cass, James, 93
Catalog cards, 18-19
Catalog of Standard Reference Materials, 143, 150
Census of Agriculture, 142
  of Business, 142
  of Construction Industries, 142
  of Governments, 142
  of Manufacturers, 142
  of Transportation, 142
Chambers's Dictionary of Science and Technology, 30
Chambers's Mineralogical Dictionary, 35 *Chambers's Shorter Six-Figure Mathematics Tables,* 145
Charts, 164
Checklist of International Business Publications, 149
CHEMCON (data base), 112, 181
Chemical Abstracts, 108, 112, 153, 180
Chemical and Engineering News, Facts and Figures, 148
Chemical and Process Engineering: Unit Operations: A Bibliographic Guide, 26
Chemical and Process Technology Encyclopedia, 44
Chemical Engineering Catalog: The Process Industries' Catalog, 139
Chemical Engineers' Handbook, 52
Chemical Industry Notes, 168
Chemical Marketing Reporter, 160

INDEX

Chemical Materials Catalog, 139
Chemical Publications: Their Nature and Use, 82
Chemical Titles, 179
Cheremisinoff, P. E., 73
Chilton, Cecil H., 52
Chow, Ven-Te, 61
Church, J. M., 48
Ciaccio, Leonard L., 73
Citation indexes, 109
Civil Engineering, 113
Civil Engineering Handbook, 53
Claire, William H., 76
Clason, W. E., 36, 39
Clauser, H. R., 46
Collections in libraries, arrangement of, 17-18
Collocott, T., 30
Commerce Clearing House, Inc., 168
Communication System Engineering Handbook, 57
Comparative Guide to Science and Engineering Programs, 93
COMPENDEX (data base), 111, 181
Comprehensive Dissertation Index, 158
Comprehensive Technical Dictionary, 38
Computer and Control Abstracts, 113
Computer Dictionary and Handbook, 54
Computer Directory and Buyers' Guide, 139
Computer Handbook, 53
Computer Literature Bibliography, 26
Computer Program Abstracts, 180
Computerized searching, 177-179
Computing Reviews, 115
Comrie, L. J., 145
Concise Chemical and Technical Dictionary, 31
Concise Dictionary of Physics and Related Subjects, 36
Concrete Construction Handbook, 55
Concrete Engineering Handbook, 55
Condensed Chemical Dictionary, 32
Condensed Computer Encyclopedia, 33

Condon, E. U., 70
Conferences, proceedings of, 129-131
  indexes to, 130-131
Considine, Douglas M., 44, 46, 56
Construction Industry Handbook, 54
Construction Inspection Handbook, 55
Contents of Contemporary Mathematical Journals and New Publications, 180
Cosmopolitan World Atlas, 165
CRC Handbook of Environmental Control, 59
  of Materials Science, 62
  of Tables for Applied Engineering Science, 144
  of Tables for Mathematics, 145
  of Tables for Probability and Statistics, 145
CRC Standard Mathematical Tables, 145
Creager, William P., 60
Crede, C. E., 77
Crocker, Sabin, 70
Crowley, Ellen T., 30
Cummins, Arthur B., 69
Cumulative Book Index, 24
Cumulative indexes, 109
Current awareness services, 177-179
Current Contents: Engineering & Technology, 180
Current Papers in Electrical & Electronics Engineering, 180
Current Programs, 130

Dana's Manual of Mineralogy, 68
Dangerous Properties of Industrial Materials, 75
Data bases, computerized, 177-179, 181
Data Book for Civil Engineers, 52
Data compilations, 143-144
  examples of, 150-151
Davidson, A., 71
Davidson, R. L., 70
Davis, Calvin V., 61
Dean, John A., 144
Dennis, William H., 86
Denti, Renzo I., 37
Design of Small Dams, 53

De Sola, Ralph, 30
de Solla Price, Derek J., 10
DeVries, Louis, 36, 37
Dictionaries, 29-41
  English language, 30-36
  foreign language, 36-41
Dictionary of Aeronautics, 39
  of Applied Geology: Mining and Engineering, 35
  of Architecture and Construction, 32
  of Astronautics, 31
  of Ceramics, 32
  of Chemical Engineering, 39
  of Chemistry and Chemical Engineering, 36
  of Chemistry and Chemical Technology in Six Languages, 40
  of Commercial Chemicals, 32
  of Computers, Automatic Control and Data Processing, 39
  of Data Processing, 33
  of Electrical Engineering: German-English, English-German, 37
  of Electrical Engineering, Telecommunications and Electronics, 40
  of Electronics and Nucleonics, 35
  of Electronics and Waveguides, 39
  of General Physics, 39
  of Industrial Digital Computer Technology, 33
  of Mechanical Engineering, 34
  of Mechanical Engineering Terms, 34
  of Metallurgy, 39
  of Mining, Minerals and Related Terms, 35
  of Modern Engineering, 37
  of Named Effects and Laws in Chemistry, Physics and Mathematics, 30
  of Nuclear Physics and Technology, 41
  of Pure and Applied Physics, 36
  of Report Series Codes, 120
  of Science and Technology, 30
  of Science and Technology (German-English and English-German), 37
  of Scientific Biography, 88
  of Telecommunication, 33
  of Television, Radar and Antennas, 39
Digital Computer User's Handbook, 54
Directories, 91-94
Directory of Engineering Societies and Related Organizations, 93
  of Engineers in Private Practice, 93
  of Information Resources in the United States: Physical Sciences, Engineering, 175
  of Published Proceedings. Series SEMT, 130
  of Registered Patent Attorneys and Agents Arranged by States and Countries, 154
  of Scientific Directories, 93
  of Testing Laboratories, Commercial-Institutional, 93
Dissertation Abstracts International. Part B—The Sciences and Engineering, 158
Dissertations, 157-158
Dobson, A., 30
Doctoral dissertations, 157-158
Dodd, A. E., 32
Dorian, A. F., 37, 40
Drawings, engineering, 164-165
Dutch, Robert A., 172
Dwight, Herbert B., 146

Edison Electric Institute, 148
Educational sources of information, 174, 175
EEE (periodical), 57
Eight-Place Tables of Trigonometric Functions for Every Second of Arc, 146
Electrical & Electronics Abstracts, 115
Electrical Engineering Design Manual, 57
Electronic Circuit Design Handbook, 57
Electronic data processing, 177-179
Electronic Engineers' Handbook, 57
Electronic Engineer's Reference Book, 58
Electronic Properties Information Center, 63

INDEX

Electronic Properties of Materials: A Guide to the Literature, 27
Electronics and Nucleonics Dictionary, 34
Electronics Buyers' Guide, 139
Electroplating Engineering Handbook, 59
Elsevier's Dictionary of Nuclear Science and Technology, 39
Emerick, Robert H., 54
E/MJ International Directory of Mining and Mineral Processing Operations, 93
Encyclopaedia of Hydraulics, Soil and Foundation Engineering, 32
of the Iron and Steel Industry, 34
Encyclopaedic Dictionary of Physics, 47
Encyclopedia of Associations, 91, 92, 173
  of Basic Materials for Plastics, 48
  of Chemical Process Equipment, 45
  of Chemical Technology, 45
  of Chemistry, 45
  of Electronics, 45
  of Engineering Materials and Processes, 46
  of Industrial Chemical Analysis, 45
  of Instrumentation and Control, 46
  of Materials Handling, 47
  of Occupational Health and Safety, 48
  of Physics, 47
  of Plastics Equipment, 47
  of Polymer Science and Technology: Plastics, Resins, Rubbers, Fibers, 47
  of the Biological Sciences, 44
  of the Chemical Elements, 45
  of Urban Planning, 48
Encyclopedias, examples of, 43-48
Energy Directory, 94
Energy Index, 115
Engineering, 86
Engineering Alloys, 68
Engineering drawings, 164-165
Engineering Index, 111
Engineering literature, how to use, 1-6, 15-19,
  types of, 10-11
Engineering Societies Library, 173
Engineers Joint Council, 88, 172
Engineers of Distinction, 88
English-German Technical and Engineering Dictionary, 37
Environment Index, 115
ERDA Energy Research Abstracts, 123

Facilities and Plant Engineering Handbook, 71
Factory Mutual System, 74
Faith, Keyes and Clark's Industrial Chemicals, 63
Fang, Hsai-Yang, 56
Ferguson, Eugene S., 86
Fink, Donald G., 56, 57
Fire Protection Handbook, 75
Fletcher, A., 146
Forsythe, William E., 148
Flugge, Wilhelm, 65
Focal Encyclopedia of Photography, 46
Foreman's Handbook, 71
Forthcoming Books, 25
Fortune, 178
Foster, Norman, 144
Foundation Engineering Handbook, 56
Fox, J., 59
French-English Dictionary for Chemists, 36
French-English Science Dictionary; for Students in Agricultural, Biological and Physical Sciences, 36
Fry, Bernard M., 81
Fundamentals Formulas of Physics, 70

Gas Engineers Handbook, 60
Gaylord, C. N., 54
Gaylord, Edwin H., Jr., 54
Gaylord, Norman G., 47
Gentle, Ernest J., 31
GEO-REF (data base), 181
German-English Dictionary for Chemists, 37
German-English Science Dictionary: For Students in Chemistry, Physics, Biology, Agriculture and Related Sciences, 36

German-English Technical and Engineering Dictionary, 37
Gibson, Eleanor B., 84
Gillispie, C. G., 88
Given, Ivan A., 69
Glass Engineering Handbook, 63
Glasstone, Samuel, 52
Glossary of Mining Geology, 35
Godfrey, Lois E., 120
Goedecke, W., 40
Goetzel, Claus G., 51
Government Reports Announcements and Index, 123, 125, 179
Governmental sources of information, 174, 175
Grabbe, Eugene M., 53
Graf, Rudolf F., 33
Graham, Arthur Kenneth, 59
Grant, E. L., 61
Grant, Julius, 32
Gray, Dwight E., 70
Gray, Peter, 44
Gray, R., 67
Grazda, Edward E., 64
Greene, J. H., 71
Griffel, William, 75
Gruenberg, Elliot L., 57
Grogan, Dennis, 80
Grossman, Fred A., 146
Guide Book to Mathematics for Technologists and Engineers, 64
Guidebook of Electronic Circuits, 58
Guide to American Scientific and Technical Directories, 94
  to Information Sources in Mining, Minerals and Geosciences, 84
  to Information Sources in Space Science and Technology, 81
  to Literature on Agricultural Engineering, 81
  to Metallurgical Information, 84
  to Reference Books, 79
  to Reference Material, 81
  to Reference Sources in the Computer Sciences, 82
  to Scientific and Technical Journals in Translation, 162
  to the Literature of Mathematics and Physics, Including Related Works on Engineering Science, 83

Guides to the literature, 79-84

Hackh's Chemical Dictionary, 32
Hammer, Willie, 75
Hampel, Clifford A., 45, 67
Hamsher, Donald H., 57
Handbook of Adhesive Bonding, 71
  of Air Conditioning, Heating and Ventilating, 51
  of Applied Hydraulics, 61
  of Applied Hydrology: A Compendium of Water-Resources Technology, 61
  of Applied Instrumentation, 56
  of Applied Mathematics, 64
  of Applied Mathematics, Selected Results and Methods, 64
  of Automation, Computation and Control, 53
  of Chemistry, 144
  of Chemistry and Physics, 141, 144
  of Chlorination: For Potable Water, Wastewater, Cooling Water, Industrial Processes, and Swimming Pools, 77
  of Electronic Materials, 63
  of Electronic Packaging, 57
  of Engineering Fundamentals, 50
  of Engineering Mechanics, 65
  of Environmental Civil Engineering, 59
  of Epoxy Resins, 72
  of Experimental Stress Analysis, 76
  of Fiberglass and Advanced Plastics Composites, 72
  of Fixture Design, 67
  of Fluid Dynamics, 61
  of Formulas for Stress and Strain, 75
  of Fuel Cell Technology, 56
  of Heat Transfer, 60
  of Heavy Construction, 55
  of High Vacuum Engineering, 77
  of Highway Engineering, 60
  of Industrial Control Computers, 53
  of Industrial Engineering and Management, 61
  of Industrial Loss Prevention, 74
  of Industrial Trade Names, 138

INDEX

of Instrumentation and Controls, 56
of Integrated Circuits, 59
of Laboratory Safety, 75
of Materials and Processes for Electronics, 58
of Mathematical Functions, with Formulas, Graphs and Mathematical Tables, 146
of Mathematical Tables and Formulas, 145
of Mechanical Specifications for Buildings and Plants: A Checkbook for Engineers and Architects, 54
of Microwave Measurements, 59
of Natural Gas Engineering, 60
of Noise Control, 69
of Ocean and Underwater Engineering, 69
of Physical Calculations, 65
of Physics, 70
of Plastics and Elastomers, 72
of Precision Engineering, 71
of Radar Measurement, 74
of Satellites and Space Vehicles, 51
of Semiconductor Electronics, 58
of System and Product Safety, 75
of Technical Writing Practices, 76
of Telemetry and Remote Control, 57
of the Engineering Sciences, 50
of X-Rays: For Diffraction, Emission, Absorption and Microscopy, 77
on Urban Planning, 76
Handbooks, 49-77
Harper, Charles A., 57, 58, 72
Harper Encyclopedia of Science, 43
Harris, Cyril M., 32, 69, 77
Harrison, Thomas J., 53
Hartman, W., 61
Hartnett, J., 60
Harvey, Anthony P, 93
Havers, John A., 55
Haviland, R. P., 51
Hawley, G. G., 32
Herrmann, T. M., 37
Hetenyi, Miklos I., 76
Hey, D. H., 45
Heyel, Carl, 71

Hicks, Tyler G., 64
Highway Research Abstracts, 113
Himmelsback, Carl J., 162
Hines, T. C., 26
Histories, 85-86
History of Civil Engineering: An Outline from Ancient to Modern Times, 86
Ho, James K. K., 88
Hochman, Stanley, 36
Holland, F. W., 58
Horger, Oscar J., 66
Horner, J. G., 34
Houghton, Bernard, 80, 83
How to Find Out about Engineering, 80
How to Find Out about Physics: A Guide to Sources of Information, 84
How to Find Out in Electrical Engineering, 83
How to Find Out in Mathematics: A Guide to Sources of Information, 83
Hoyt, Samuel S., 66
Huckert, Jesse, 147
Hughes, L. E. C., 35, 58
Hundred Years of Metallurgy, 86
Hunter, Lloyd P., 58
Hurlbut, Cornelius S. Jr., 68
Huskey, Harry D., 53
Hutchison, J. W., 150
Hydroelectric Handbook, 60
Hyslop, Marjorie R., 84

IES Lighting Handbook: The Standard Lighting Guide, 62
IFAC Multilingual Dictionary of Automatic Control Terminology, 38
Illuminating Engineering Society, 62
IMM Abstracts, 117
Index of Federal Specifications and Standards, 135
  of Mathematical Tables, 146
  of Specifications and Standards, 135
Index to Foreign Market Reports, 143
  to Scientific Reviews, 99
Indexing services, 107-109
Indicative abstracts, 24, 107

Industrial Chemicals, 63
Industrial Engineering Handbook, 62
Industrial Pollution Control Handbook, 73
Industrial Research Laboratories of the United States, 94
Industrial sources of information, 174, 175
Information analysis centers, role of, 14-19
Informative abstracts, 24, 107
Ingenious Mechanisms for Designers and Inventors, 65
INIS Atomindex, 122, 123
INSPEC (data base), 181
International Abstracts in Operations Research, 116
International Academy of Astronautics, 40
International Aerospace Abstracts, 112, 124
International Critical Tables of Numerical Data; Physics, Chemistry and Technology, 143, 150
International Dictionary of Applied Mathematics, 34
International Dictionary of Metallurgy, Mineralogy, Geology and the Mining and Oil Industries, 38
International Dictionary of Physics and Electronics, 35
International Labour Organization, 48
International Maps and Atlases in Print, 165
Invisible college, 173
Ireson, William Grant, 61, 74
Irregular Serials and Annuals, 130
ISA Handbook of Control Valves, 150
Italian-English, English-Italian Technical Dictionary, 37

Jahnke, Eugene, 146
James, Glenn, 40
James, R. C., 40
Jane's All the World's Aircraft, 95
Jane's World Mining—Who Owns Whom, the World Companion to Mining Investment, 139
Janes, M. F., 55
Jenkins, Frances B., 25
Johnson, H. Thayne, 27
Jones, Franklin D., 65
Jordain, Philip B., 33
Jordan, Stello, 76
Journal of Physical and Chemical Reference Data, 150
Juran, Joseph M., 74
Justin, J. D., 60

Kaelble, E. F., 77
Kallen, Howard P., 56
Kaplan, Stuart R., 84
Katz, Donald L., 60
Kaye, George W., 147
Kempe's Engineers Year-Book, 50
Kennan, Joseph H., 148
Kent, R. T., 65
Keystone Coal Industry Manual, 139
King, Reno C., 70
King, Richard L., 27
Kingzett's Chemical Encyclopaedia: A Digest of Chemistry and Its Industrial Applications, 45
Kirk, R. E., 45
Klein, Bernard, 94
Klerer, Melvin, 54, 146
Koral, R. L., 51
Korn, Granino A., 53, 54, 64
Korn, T. M., 64
Kozlov, B. A., 74
KWIC indexes, 109
KWOC indexes, 109
Kyed, James M., 94

Laby, T. H., 147
La Londe, William S., 55
Lange's Handbook of Chemistry, 144
Lapedes, Daniel N., 31
Lasworth, Earl James, 80
Lavine, I., 138
Lee, Henry, 72
Lenk, John D., 150
Leslie, W. H. P., 67
Lewis, Bernard T., 71
Libraries, card catalogs in, 1-2, 16-19
  collection arrangement in, 17-18
  how to use, 1-6, 15-19
  networks of, 19
  role of, 13-19
Literature, growth of, 10

types of, 10-11
Literature of Chemical Technology, 82
Literature searching, basic rules of, 3-6
  computerized, 177-179
  retrospective, 178-179
Lovett, D. R., 30
Lowenheim, Frederick A., 63
Lubin, George, 72
Lund, Herbert F., 73
Lynch, Charles T., 62

McDowell, C. H., 34
McGraw-Hill Basic Bibliography of Science and Technology, 26
McGraw-Hill Dictionary of Scientific and Technical Terms, 31
McGraw-Hill Encyclopedia of Environmental Science, 46
McGraw-Hill Encyclopedia of Science and Technology, 44
McGraw-Hill Encyclopedia of Space, 44
McGraw-Hill Modern Men of Science, 88
McGraw-Hill Yearbook of Science and Technology, 44, 95
Machinery's Handbook: A Reference Book for the Mechanical Engineer, Draftsman, Toolmaker and Machinist, 68
Machol, Robert E., 62
MacRae's Blue Book, 138
Magnetic tape services, 177-181
Maintenance Engineering Handbook, 72
Malinowsky, H. Robert, 80
Man and the Environment: A Bibliography of Selected Publications of the United Nations System, 1946-1971, 27
Management Information Systems Handbook, 61
Manual of Classification, 154
Manual of Steel Construction, 54
Manufacturers, directories of, 138-140
Manufacturers' catalogs, 137-140
Manufacturing Planning and Estimating Handbook, 71
Map Collections in the United States and Canada, 164, 165
Maps, 163-164

indexes to, 165-166
MARC (Machine Readable Catalog), 19
Mark, Herman F., 47
Marks, L., 65
Marks, Robert W., 31
Markus, John, 34, 58
Marron, J. P., 71
Master's essays, indexes to, 158
Masters Theses in the Pure and Applied Sciences Accepted by Colleges and Universities in the United States and Canada, 158
Matarazzo, James M., 94
Materials Data Book for Engineers and Scientists, 63
Materials Handbook: An Encyclopedia for Purchasing Agents, Engineers, Executives and Foremen, 46
Mathematics Dictionary, 40
Mathematical Handbook for Scientists and Engineers, 64
Mathematical Handbook of Formulas and Tables, 146
Mathematical Reviews, 116
Mathematical Tables and Other Aids to Computation, 141
Mathematical Tables of Elementary and Some Higher Mathematical Functions, 146
Mathematics of Computation, 141
Maynard, H. B., 62
Maynard, Jeff, 33
Mead, William J., 45
Mechanical Design and Systems Handbook, 66
Mechanical Design: Reference Sources, 83
Mechanical Engineering: The Sources of Information, 83
Mechanical Engineers' Handbook, 65
MEDLARS (data base), 181
Mellon, M. G., 82
Menzel, Donald H., 70
Merritt, Frederick S., 52, 55
METADEX (data base), 181
Metal Finishing Guidebook Directory, 139
Metal Finishing, 140
Metal Statistics, 148
Metallic Materials, 68
Metallurgical Abstracts, 117

Metals Abstracts, 117
Metals Abstracts Index, 117
Metals Handbook, 66
Metals Reference Book, 147
Microfilm copies of technical reports, 120
Minerals Yearbook, 95
Mining Engineers' Handbook, 68
Mining Year Book, 140
Modern Dictionary of Electronics, 33
Modern Plastics, 47
Modern Plastics Encyclopedia, 47
Mohr, J. Gilbert, 73
Mohrhardt, F. E., 81
Monthly Catalog of United States Government Publications, 123, 124, 142
Moody's Investors Service, Inc., 168
Moran, Marguerite A., 63
Morrow, L. C., 72
Moser, Reta C., 31
Moss, J. B., 63
Multilingual dictionaries, 38-41
Myers, John J., 69

National Atlas of the United States of America, 165
National Electrical Code, 57
National Federation of Science Abstracting and Indexing Services, 108
National Fire Protection Association, 57, 75
National Referral Center, 175
National Research Council, 150
National Standard Reference Data System, 143, 150
Nation's Business, 178
Nautical Dictionary, 40
Nayler, G. H., 34
Nayler, Joseph L., 31, 34
Neidhardt, P., 40
Nelson, A., 35
Nelson, K. D., 35
Neville, Kris, 72
New Acronyms and Initialisms, 30
New Dictionary and Handbook of Aerospace: With Special Sections on the Moon and Lunar Flight, 31
New Research Centers, 92
New Serial Titles: A Union List of Serials Commencing Publication after Dec. 31, 1949, 105
New Table of Indefinite Integrals, Computer Processed, 146
New Technical Books: A Selective List with Descriptive Annotations, 26
New York Times Index, 160
Newman, James R., 43
Newspapers, 159-160
1967 ASME Steam Tables, 148
NTIS (data base), 181
Nuclear Science Abstracts, 122, 123, 124
Numerical Control User's Handbook, 67

Oberg, Erik, 68
O'Brien, James J., 55
Oceanic Abstracts, 117
Oceanic Citation Journal, 117
Oceanic Index, 117
Oceans: Their Physics, Chemistry and General Biology, 69
O'Connor, James J., 62
Odishaw, H., 70
Official Gazette: Patents, 154
Official Gazette: Trademarks, 155
Ohio College Library Center, 19
Oil, Paint and Drug Reporter, 160
Oleesky, Samuel F., 73
Operations Research Management Science, 116
Oppermann, Alfred, 37
Original Roget's Thesaurus of English Words and Phrases, 172
Osborne, A. K., 33
Osenton, J., 40
Othmer, D. F., 45
Oxide Handbook, 63

Palmer, Archie M., 92
Paperbound Books in Print, 25
Parke, Nathan Grier III, 83
Parker, Earl R., 63
Parsons, Stanley Alfred James, 80
Patents, 153-155
Patterson, Austin M., 36, 37
Pearson, Carl E., 64
Peele, R., 68
Pemberton, John E., 83
Periodical indexing and abstracting services, 107-118

INDEX

Periodicals, 103-118
  directories of, 104-106
  union lists of, 105-106
Permuterm Subject Index, 111
Perry, John H., 52
Personal sources of information, 173, 175
Peters, Jean, 146
Petroleum Processing Handbook, 70
Petroleum Transportation Handbook, 69
Physics Abstracts, 117
Physics Handbook, 70
Physics Literature: A Reference Manual, 84
Piping Handbook, 70
Plant Engineering Handbook, 72
Plastics Engineering Handbook, 73
Plumb, P., 83
Pollution Abstracts, 118
Pollution Engineering Practice Handbook, 73
Poor's Register of Corporations, Directors and Executives, 169
Potter, James H., 50
Practical Semiconductor Databook for Electronic Engineers and Technicians, 150
Practical Tables for Building Construction, 144
Prentice-Hall, Inc., 169
Proceedings in Print, 131
Proceedings of symposia and meetings, 129-131
Production Handbook, 72
Production and Inventory Control Handbook, 71
Professional societies, information from, 173, 175
Progress in ..... series, 99
Properties of Engineering Materials, 63
Publishers Trade List Annual, 24
Publishers Weekly, 25
Purdue Universitiy. Thermophysical Properties Research Center, 147

Quality Control and Applied Statistics, 116
Quality Control Handbook, 74
Quotations, 171-172

Radar Handbook, 74
Radome Engineering Handbook: Design and Principles, 74
RAND Abstracts, 124
Rapport, Samuel, 86
Rare Metals Handbook, 67
Reactor Handbook, 52
Redman, Helen F., 120
Reference Data for Radio Engineers, 58
Reference materials in libraries, 2
Reference Sources in Science and Technology, 80
Reithmaier, Lawrence W., 31
Reliability Handbook, 74
Reports, see Technical Reports
Research Centers Directory, 92
Review of Metal Literature, 117
Review series, 97-99
Richey, C. B., 50
Roberts, Elizabeth P., 81
Rohsenow, W. M., 60
Ross, Robert B., 68
Ross, S. D., 56
Rothbart, Harold P., 66
Russian-English Chemical and Polytechnical Dictionary, 38
Russian-English Scientific and Technical Dictionary,

SAE Handbook, 76
Sale of New One-Family Homes, 142
Samsonov, G. V., 63
Satellite photographs, 164
Sax, N. Irving, 75
Say, Maurice G., 57
Schneider, John H., 180
Schwenkhagen, H. F., 37
Science Abstracts, 113, 115, 117
Science and Engineering Reference Sources: A Guide for Students and Librarians, 80
Science and Technology: An Introduction to the Literature, 80
Science Citation Index, 111
Science Reference Sources, 25
Scientific and Technical Aerospace Reports, 112, 122, 124
Scientific Meetings, 131
Scientific, Technical and Engineering Societies Publications in Print 1974-1975, 94

Scientific, Technical and Related Societies of the United States, 94
SCISEARCH (data base), 181
SDI services, 178-179
Searching of literature, basic rules of, 3-6
  computerized, 177-179
  retrospective, 178-179
Searching the Chemical Literature, 6
Security aspects of technical reports, 119-120
Seelye, Elwyn E., 52
Segditsas, P. E., 40
Selby, Samuel M., 145
Selected Publications to Aid Domestic Business and Industry, 149
Selected Rand Abstracts, 124
Selective dissemination of information, 178-179
Sell, Lewis L., 38
Semendyayev, K. A., 64
Shand, E. B., 63
Sheehy, Eugene P., 79
Shiers, George, 86
Shock and Vibration Handbook, 77
Short Dictionary of Mathematics, 34
Siddall, James N., 83
Simonds, Herbert R., 47, 48
Singer, T. E., 82
Sippl, Charles J., 54
Skolnik, Merrill I., 74
SME Mining Engineering Handbook, 69
Smith, Julian F., 82
Smithells, Colin J., 147
Smithsonian Physical Tables, 148
Smithsonian Science Information Exchange (SSIE), 179
Snell, C. T., 32
Snell, Foster D., 32, 45
Sobecka, Z., 40
Societies, information from, 173, 175
Society of Automotive Engineers, 76
Society of the Plastics Industry, Inc., 73
Sorensen, K. E., 61
Souders, Mott, 50
Sourcebook of Electronic Circuits, 58

Sourcebook on Atomic Energy, 52
Sources of Information on Atomic Energy, 81
Space-Age Acronyms: Abbreviations and Designations, 31
Space Materials Handbook, 51
Special Libraries Association, Geography and Map Division. Directory Revision Committee, 165
Specifications, 133-135
SPI Handbook of Technology and Engineering of Reinforced Plastics/Composites, 73
Spiegel, Murray R., 146
SPSE Handbook of Photographic Science and Engineering, 70
SSIE (data base), 181
SSIE Science Newsletter, 179
Standard and Poor's, 169
Standard Dictionary of Computers and Information Processing, 33
Standard Handbook for Civil Engineers, 52
  for Electrical Engineers, 56
  for Mechanical Engineers, 65
  of Engineering Calculations, 64
  of Lubrication Engineering, 62
Standards, 133-135
Standards and Specifications Information Sources, 135
Staniar, William, 72
Statistical Abstract of the United States, 142, 149
Statistical Yearbook, 142, 149
Statistical Year Book of the Electric Utility Industy, 148
Statistics, 142-143, 148-149
Statistics Sources, 149
Steam Tables; Thermodynamic Properties of Water Including Vapor, Liquid and Solid Phases, 148
Steere, Norman V., 75
Steinherz, H. A., 77
Stratton, G. B., 105
Straub, Conrad P., 59
Straub, Hans, 86
Streeter, Victor L., 61
Strock, Clifford, 51
Structural Engineering Handbook, 54
Structural Steel Designers' Handbook, 55

INDEX

Struglia, Erasmus J., 135
Stubbs, F. W., Jr., 55
Sube, R., 41
Subject Guide to Books in Print, 24, 25
Subject Guide to Forthcoming Books, 25
Sucher, Max, 59
Survey of Commercially Available Computer-Readable Bibliographic Data Bases, 180
Susskind, Charles, 45
Sverdrup, H. U., 69
Sweet's Catalog File, 137, 140
Symposia, 129-131
   proceedings of, 129-131
System Engineering Handbook, 62

Tables, 141, 144-148
Tables of Higher Functions, 146
Tables of Physical and Chemical Constants and Some Mathematical Functions, 147
Tapia, E. W., 84
Technical Abstract Bulletin, 122, 124
Technical Book Review Index, 26
Technical Dictionary of Data Processing, Computers and Office Machines, 38
Technical Dictionary of Electronics, 40
Technical Information Sources: A Guide to the Patents, Standards and Technical Reports Literature, 80
Technical report indexing and abstracting services, 121-125
Technical reports, 119-125
  citations of, 120
  indexes and abstracts for, 121-125
  microfilm versions of, 120
  security aspects of, 119-120
  sources of, 119
Techniques of Metals Research, 67
Telecommunication Dictionary, 41
Thermal Conductivity-Metallic Elements and Alloys, 147
Thesauri, 171-172
Thesaurus of Engineering and Scientific Terms, 172
Thewlis, James, 36, 47

Thomas, Harry E., 59
Thomas, Robert C., 30
Thomas, Woodlief Jr., 70
Thomas Register of American Manufacturers and Thomas Register Catalog File, 138
Thrush, Paul W., 35
Timber Construction Manual, 76
Timber Design and Construction Handbook, 76
Timber Engineering Company, 76
Tools Engineers Handbook, 67
Touloukian, Y. S., 147
Trade catalogs, 137-140
Trademarks, indexes to, 154-155
Translation Register-Index, 161, 162
Translations, 161-162
Tuma, Jan J., 65
Tuve, G. L., 144

Ulrich's International Periodicals Directory, 105
Union List of Serials in Libraries of the U.S. and Canada, 105
United Engineering Center, 173
Urquhart, Leonard C., 53
U.S. Atomic Energy Commission, 52
U.S. Bureau of Mines, 95
U.S. Department of Commerce, 149
U.S. Department of Defense, 135
U.S. Department of the Interior, 35
U.S. Department of the Interior. Bureau of Reclamation, 53
U.S. Energy Research and Development Administration, 135
U.S. General Services Administration, 135
U.S. Government Research and Development Reports, 123
U.S. Library of Congress. National Referral Center, 175
U.S. National Bureau of Standards, 26, 146
U.S. National Bureau of Standards. Office of Standard Reference Materials, 150
U.S. Patent Office, 154, 155
Use of Biological Literature, 81
Use of Chemical Literature, 82
Ushakov, I. A., 74
Using the Chemical Literature, 82

Van Horn, Kent, 68
Van Nostrand's Scientific Encyclopedia, 44
Visser, A., 41
Vollmer, Ernst, 32

VSMF Design Engineering System, 138

Waddell, Joseph J., 55
Walford, A. J., 81
Wall Street Journal, 159, 160, 167
Walton, J. D. Jr., 74
Ward, H. R., 74
Wasserman, Paul, 149
Water and Water Pollution Handbook, 73
Water Pollution Abstracts, 118
Water Quality and Treatment: a Handbook of Public Water Supplies, 77
Weekly Government Abstracts, 123
Weekly Record, 25
Weik, Martin H., 33
Welding Handbook, 67
Westman, H. P., 58
White, George C., 77
White, G. K., 34
Whitford, Robert N., 84
Whittick, Arnold, 48
Who Is Publishing in Science: An International Directory of Research and Development Scientists, 89
Who's Who in Computers and Data Processing, 89
Who's Who in Engineering, 89
Who's Who in Science in Europe, 89
Wilson, William K., 94
Winchell, Constance M., 79
Winterkorn, Hans F., 56
Winton, Harry N. M., 27
Woldman, Norman E., 68
Woodburn, Henry M., 82
Woodley, Douglas R., 47
World Directory of Environmental Research Centers, 94
World Guide to Scientific Associations, 92
World Index of Scientific Translations and List of Translations Notified to ETC, 162
World List of Scientific Periodicals, Published in the Years 1900-1960, 103, 105
World Mining Glossary of Mining, Processing and Geological Terms, 41
World of Learning, 91, 92
World Who's Who in Science: A Biographical Dictionary of Notable Scientists from Antiquity to the Present, 89
Wright, Helen, 86
Wyatt, H. V., 81
Wyllie, R. J. M., 41

Yates, Bryan, 84
Yearbooks, 91-92
  examples of, 95
Youden, W. W., 27
Young, Richard A., 73

Zilly, Robert G., 59
Zimmerman, O. T., 138

Ref
T
10.7
M68
1976

MAR 18 1977